Applied Mathematical Sciences

Volume 175

The mathematization of all sciences, the fading of traditional scientific boundaries, the impact of computer technology, the growing importance of computer modeling and the necessity of scientific planning all create the need both in education and research for books that are introductory to and abreast of these developments. The purpose of this series is to provide such books, suitable for the user of mathematics, the mathematician interested in applications, and the student scientist. In particular, this series will provide an outlet for topics of immediate interest because of the novelty of its treatment of an application or of mathematics being applied or lying close to applications. These books should be accessible to readers versed in mathematics or science and engineering, and will feature a lively tutorial style, a focus on topics of current interest, and present clear exposition of broad appeal. A compliment to the Applied Mathematical Sciences series is the Texts in Applied Mathematics series, which publishes textbooks suitable for advanced undergraduate and beginning graduate courses.

More information about this series at https://link.springer.com/bookseries/34

Rafael Martí · Gerhard Reinelt

Exact and Heuristic Methods in Combinatorial Optimization

A Study on the Linear Ordering and the Maximum Diversity Problem

Second Edition

Rafael Martí
Departamento de Estadística e
Investigación Operativa
Universitat de València
Valencia, Spain

Gerhard Reinelt
Department of Computer Science
University of Heidelberg
Heidelberg, Germany

ISSN 0066-5452   ISSN 2196-968X   (electronic)
Applied Mathematical Sciences
ISBN 978-3-662-64879-7   ISBN 978-3-662-64877-3   (eBook)
https://doi.org/10.1007/978-3-662-64877-3

Cover design: deblik, Berlin

This Springer imprint is published by the registered company Springer-Verlag GmbH, DE part of Springer Nature.
The registered company address is: Heidelberger Platz 3, 14197 Berlin, Germany

To Mariola for her life, and to Mila for giving life to her.

   Rafa Martí

Preface

Faced with the challenge of solving hard optimization problems that abound in the real world, classical methods often encounter serious difficulties. Important applications in business, engineering or economics cannot be tackled by the solution methods that have been the predominant focus of academic research throughout the past four decades. Exact and heuristic approaches are dramatically changing our ability to solve problems of practical significance and are extending the frontier of problems that can be handled effectively. In this text we describe state-of-the-art optimization methods, both exact and heuristic, for two difficult optimization problems: the linear ordering problem (LOP), and the maximum diversity problem (MDP). In this way, we provide the reader with the background, elements and strategies to tackle a wide range of different combinatorial optimization problems.

The idea for writing the first edition of this book came up when the authors met at the University of Valencia in 2005. While comparing our experiences with regard to various aspects of the LOP, we realized that most of the optimization technologies had been successfully applied to solve this problem. We also found that there were only a small number of books covering all state-of-the-art optimization methods for hard optimization problems (especially considering both exact methods and heuristics together). We thought that the LOP would make an ideal example to survey these methods applied to one problem and felt the time was ripe to embark on the project of writing a monograph.

In this second edition, we performed a major revision of the book, to offer the reader a wider approach to combinatorial optimization. In particular, we included a second problem, the maximum diversity problem (MDP), to describe the optimization methods covered in the book both in the LOP and in the MDP. Considering that in the LOP solutions are usually represented with permutations, and in the MDP with binary variables, the adaptation of the different optimization technologies to these two problems, provide readers with more elements, tools, and search strategies to create their own solving methods

This textbook is devoted to the LOP and the MDP, their origins, applications, instances and especially to methods for their effective approximate or exact solution. Our intention is to provide basic principles and fundamental ideas and reflect the

state-of-the-art of heuristic and exact methods, thus allowing the reader to create his
or her personal successful applications of the solution methods. The book is meant
to be of interest for researchers and practitioners in computer science, mathematics,
operations research, management science, industrial engineering, and economics. It
can be used as a textbook on issues of practical optimization in a master's course or
as a reference resource for engineering optimization algorithms.

To make the book accessible to a wider audience, it is to a large extent self-
contained, providing the reader with the basic definitions and concepts in optimization. However, in order to limit the size of this monograph we have not included
extensive introductions. Readers interested in further details are referred to appropriate textbooks such as [4, 103, 129, 155, 156, 163].

The structure of this book is as follows. Chapter 1 provides an introduction to the
two problems and their applications, and describes the sets of benchmark instances
which we are using for our computational experiments and which have been made
publically available. Chapter 2 describes basic heuristic methods such as construction and local searches. Chapter 3 expands on Chapter 2 and covers meta-heuristics
in which the simple methods are now embedded in complex solution algorithms
based on different paradigms, such as evolution or learning strategies. Chapter 4
discusses branch-and-bound, the principal approach for solving difficult problems
to optimality. A special version based on polyhedral combinatorics, branch-and-cut,
is presented in Chapter 5. Chapter 6 deals in more detail with the linear ordering
polytope which is at the core of branch-and-cut algorithms. The book concludes
with Chapter 7, where a number of further aspects of the LOP and MDP, and potential issues for further research are described.

Rafael Martí's research was partially supported by the Ministerio de Ciencia e
Innovación of Spain (Grant Ref. PGC2018-0953322-B-C21 /MCIU/AEI/FEDER-
UE).

We are in debt to many people, but in particular to some very good friends and
colleagues who helped us to gain a deeper understanding of our two problems: Vi-
cente Campos, Thomas Christof, Angel Corberán, Abraham Duarte, Fred Glover,
Martin Grötschel, Michael Jünger, Manuel Laguna, Anna Martínez-Gavara, Fran-
cisco Parreño, and Mauricio Resende.

Valencia, Heidelberg,   *Rafael Martí*
September 2021   *Gerhard Reinelt*

Contents

Contents

Chapter 1
Introduction

Abstract This chapter introduces the basic definitions, main elements, applications, and instances of two optimization problems, the linear ordering problem (LOP), and the maximum diversity problem (MDP). We will use them in the next chapters to describe heuristics, meta-heuristics and exact approaches, and to report our experiments.

The LOP is one of the classical combinatorial optimization problems which was already classified as NP-hard in 1979 by Garey and Johnson [64]. It has received considerable attention in various application areas ranging from archeology and scheduling to economics. Solution methods for the LOP have been proposed since 1958, when Chenery and Watanabe outlined some ideas on how to obtain solutions for this problem. The interest in this problem has continued over the years, resulting in the book [143] and many recent papers in scientific journals. This chapter surveys the main LOP applications and benchmark library of instances LOLIB.

The challenge of maximizing the diversity of a collection of points arises in a variety of settings, from location to genetics. The growing interest of dealing with diversity translated into mathematical models and computer algorithms in the late eighties, when Michael Kuby studied dispersion maximization in general graphs [54]. The MDP is the first model proposed and the most studied to deal with diversity, and was classified as NP-hard. Many optimization methods have been proposed to obtain efficient solutions to this problem, which makes it especially convenient as an illustrative example in this book. This chapter surveys the different models proposed to maximize diversity, their applications, and the benchmark library of instances MDPLIB.

1.1 The Linear Ordering Problem

In its graph version the LOP is defined as follows. Let $D_n = (V_n, A_n)$ denote the complete digraph on n nodes, i.e., the directed graph with node set $V_n = \{1, 2, \ldots, n\}$ and the property that for every pair of nodes i and j there is an arc (i, j) from i to j and

R. Martí and G. Reinelt, *Exact and Heuristic Methods in Combinatorial Optimization*, Applied Mathematical Sciences 175, https://doi.org/10.1007/978-3-662-64877-3_1

an arc (j,i) from j to i. A *tournament* (or *spanning tournament*) T in A_n consists of a subset of arcs containing for every pair of nodes i and j either arc (i,j) or arc (j,i), but not both. A *(spanning) acyclic tournament* is a tournament without directed cycles, i.e., not containing an arc set of the form $\{(v_1,v_2),(v_2,v_3),\ldots,(v_k,v_1)\}$ for some $k > 1$ and distinct nodes $v_1,v_2,\ldots,v_k$.

A *linear ordering* of the nodes $\{1,2,\ldots,n\}$ is a ranking of the nodes given as linear sequence, or equivalently, as a permutation of the nodes. We denote the linear ordering that ranks node v_1 first, v_2 second, etc., and v_n last by $\langle v_1,v_2\ldots,v_n\rangle$ and write $v_i \prec v_j$ if node v_i is ranked before node v_j. If σ denotes a linear ordering, then $\sigma(i)$ gives the position of node i in this ordering. We will also consider *partial orderings* where only a subset of the nodes is ranked or only some pairs are compared.

It is easy to see that an acyclic tournament T in A_n corresponds to a linear ordering of the nodes of V_n and vice versa: the node ranked first is the one without entering arcs in T, the node ranked second is the one with one entering arc (namely from the node ranked first), etc., and the node ranked last is the one without leaving arcs in T.

Usually, ordering relations are weighted and we have weights c_{ij} giving the benefit or cost resulting when node i is ranked before node j or, equivalently, when the arc (i,j) is contained in the acyclic tournament. The (weighted) *linear ordering problem* is defined as follows.

Linear ordering problem

Given the complete directed graph $D_n = (V_n, A_n)$ with arc weights c_{ij} for every pair $i, j \in V_n$, compute a spanning acyclic tournament T in A_n such that $\sum_{(i,j)\in T} c_{ij}$ is as large as possible.

Alternatively, the LOP can be defined as a matrix problem, the so-called *triangulation problem*.

Triangulation problem

Let an (n,n)-matrix $H = (H_{ij})$ be given. Determine a simultaneous permutation of the rows and columns of H such that the sum of superdiagonal entries becomes as large as possible.

Obviously, by setting arc weights $c_{ij} = H_{ij}$ for the complete digraph D_n, the triangulation problem for H can be solved as a linear ordering problem in D_n. Conversely, a linear odering problem for D_n can be transformed to a triangulation problem for an (n,n)-matrix H by setting $H_{ij} = c_{ij}$ and the diagonal entries $H_{ii} = 0$.

Consider as an example the $(5,5)$-matrix

$$H = \begin{pmatrix} 0 & 16 & 11 & 15 & 7 \\ 21 & 0 & 14 & 15 & 9 \\ 26 & 23 & 0 & 26 & 12 \\ 22 & 22 & 11 & 0 & 13 \\ 30 & 28 & 25 & 24 & 0 \end{pmatrix}.$$

The sum of its superdiagonal elements is 138. An optimum triangulation is obtained if the original numbering $(1,2,3,4,5)$ of the rows and columns is changed to $(5,3,4,2,1)$, i.e., the original element H_{12} becomes element $H_{\sigma(1)\sigma(2)} = \tilde{H}_{54}$ in the permuted matrix. Thus the optimal triangulation of H is

$$\tilde{H} = \begin{pmatrix} 0 & 25 & 24 & 28 & 30 \\ 12 & 0 & 26 & 23 & 26 \\ 13 & 11 & 0 & 22 & 22 \\ 9 & 14 & 15 & 0 & 21 \\ 7 & 11 & 15 & 16 & 0 \end{pmatrix}.$$

Now the sum of superdiagonal elements is 247.

1.1.1 Applications

We review some of the many applications of the linear ordering problem.

Equivalent Graph Problems

The *acyclic subdigraph problem* (ASP) is defined as follows. Given a directed graph $D = (V,A)$ with arc weights d_{ij}, for all $(i,j) \in A$, determine a subset $B \subseteq A$ which contains no directed cycles and has maximum weight $d(B) = \sum_{(i,j) \in B} d_{ij}$.

It can easily be seen that this problem is equivalent to the LOP. For a given ASP define a LOP on D_n, where $n = |V|$, by setting for $1 \leq i, j \leq n, i \neq j$:

$$c_{ij} = \begin{cases} \max\{0, d_{ij}\}, & \text{if } (i,j) \in A, \\ 0, & \text{otherwise.} \end{cases}$$

If T is a tournament of maximum weight, then $B = \{(i,j) \in T \cap A \mid c_{ij} > 0\}$ is an acyclic subdigraph of D of maximum weight. In the opposite direction, by adding a suitably large constant, we can transform a given LOP into an equivalent one where all weights are strictly positive. Then an acyclic subdigraph of maximum weight is a tournament.

The *feedback arc set problem* (FBAP) in a weighted digraph $D = (V,A)$ consists of finding an arc set B of minimum weight such that $A \setminus B$ is acyclic, i.e., such that B is a so-called *feedback arc set* intersecting every dicycle of D. Obviously, FBAP and ASP are equivalent because they are complementary questions.

Fig. 1.1 shows a digraph on 9 nodes where the arcs of a minimum feedback arc
set are drawn as dotted lines. If the six arcs of the feedback arc set are removed, we
obtain an acyclic arc set.

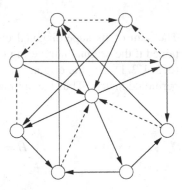

Fig. 1.1 A digraph with
minimum feedback arc set

Related Graph Problems

There are some further problems dealing with acyclic subdigraphs. The *node induced acyclic subdigraph problem* asks for a node set $W \subseteq V$ such that the subdigraph $(W, A(W))$ is acyclic. (Here $A(W)$ denotes the set of arcs with both end nodes
in W.) The problem can be defined either with node weights d, and $d(W)$ is to be
maximized, or with arc weights c where $c(A(W))$ has to be maximum. Analogously,
the *feedback node set problem* is to find a set $W \subseteq V$ such that $(V \setminus W, A(V \setminus W))$ is
acyclic. Here, sums of node weights or arc weights have to be minimized.

The request that solution digraphs have to be node induced adds a further complexity. These problems cannot be transformed to a pure linear ordering problem
and are even more difficult.

Aggregation of Individual Preferences

Linear ordering problems may occur whenever rankings of some objects are to
be determined. Consider for example the following situation. A set of n objects
$O_1, O_2, \ldots, O_n$ is given which have to be rated by m persons according to their individual preferences. Then a ranking of these objects is to be found which reflects
these single rankings as closely as possible. The first question to be answered is
how the individual rankings can be obtained. One solution is a pairwise comparison
experiment. For any pair O_i and O_j, $1 \le i < j \le n$, of objects each person decides
whether O_i should be preferred to O_j or vice versa. The results of these $m\binom{n}{2}$ comparisons are stored in an (n, n)-matrix $H = (H_{ij})$ where $H_{ij} = $ number of persons
preferring object O_i to object O_j. A ranking of these objects which infers as few

contradictions to the individual rankings as possible can be obtained by triangulating H. It should be remarked that there are various statistical methods to aggregate single preference relations to one relation.

This area of application is the oldest one of the LOP. In 1959 Kemeny [96] posed the following problem (*Kemeny's problem*). Suppose that there are m persons and each person i, $i = 1, \dots, m$, has ranked n objects by giving a linear ordering T_i of the objects. Which common linear ordering aggregates the individual orderings in the best possible way? We can solve this problem as a linear ordering problem by setting c_{ij} = number of persons preferring object O_i to object O_j. Note that this is basically the problem stated above, but this time the relative ranking of the objects by each single person is consistent (which is not assumed above).

Slater [157], in 1961, asked for the minimum number of arcs that have to be reversed to convert a given tournament T into an acyclic tournament. In the context of preferences, the input now is a ranking for all pairs i and j of objects stating whether i should be preferred to j or vice versa and the problem is to find a the maximum number of pairwise rankings without contradiction. Also *Slater's problem* can also be solved as a LOP, namely by setting

$$c_{ij} = \begin{cases} 1, & \text{if } (i,j) \in T, \\ 0, & \text{otherwise.} \end{cases}$$

Questions of this type naturally occur in the context of voting (How should a fair distribution of seats to parties be computed from the votes of the electors?) and have already been studied in the 18th century by Condorcet [42].

Binary Choice Probabilities

Let S_n denote the set of all permutations of $\{1,2,\dots,n\}$ and let P be a probability distribution on S_n.

Define the *induced (binary choice) probability system* p for $\{1,2,\dots,n\}$ as the mapping $p : \{1,2\dots,n\} \times \{1,2\dots,n\} \setminus \{(i,i) \mid i = 1,2,\dots,n\} \to [0,1]$ where

$$p(i,j) = \sum_{S \in S_n,\, i \prec j \text{ in } S} P(S).$$

The question of whether a given vector p is a vector of binary choice probabilities according to this definition is of great importance in mathematical psychology and the theory of social choice (see [60] for a survey).

In fact, the set of binary choice vectors is exactly the linear ordering polytope which will play a prominent role later in this book.

Triangulation of Input-Output Tables

One field of practical importance in economics is *input-output analysis*. It was pioneered by Leontief [108, 109] who was awarded the Nobel Prize in 1973 for his fundamental achievements. The central component of input-output analysis is the so-called *input-output table* which represents the dependencies between the different branches of an economy. To make up an input-output table the economy of a country is divided into sectors, each representing a special branch of the economy. An input-output table shows the transactions between the single sectors in a certain year. To be comparable with each other all amounts are given in monetary values. Input-output analysis is used for forecasting the development of industries and for structural planning (see [88] for an introductory survey).

Triangulation is a means for a descriptive analysis of the transactions between the sectors. In a simple model of production structure the flow of goods begins in sectors producing raw material, then sectors of manufacturing follow, and in the last stage goods for consumption and investments are produced. A real economy, of course, does not show such a strict linearity in the interindustrial connections, here there are flows between almost any sectors. Nevertheless it can be observed that the main stream of flows indeed goes from primary stage sectors via the manufacturing sectors to the sectors of final demand. Triangulation is a method for determining a hierarchy of all sectors such that the amount of flow incompatible with this hierarchy (i.e., from sectors ranked lower to sectors ranked higher) is as small as possible. Such rankings allow interpretations of the industrial structure of a country and comparisons between different countries.

Optimal Weighted Ancestry Relationships

This application from anthropology has been published in [73]. Consider a cemetery consisting of many individual gravesites. Every gravesite contains artifacts made of different pottery types. As gravesites sink over the years and are reused, it is a reasonable assumption that the depth of a pottery type is related to its age. So every gravesite gives a partial ordering of the pottery types contained in it. These partial orderings may not be consistent in the sense that pairs of pottery types may be ranked differently depending on the gravesite. The task of computing a global ordering with as few contradictions as possible amounts to solving a linear ordering problem in the complete directed graph where the nodes correspond to the pottery types and the arc weights are aggregations of the individual partial orderings. In [73] several possibilities for assigning arc weights are discussed and a simple heuristic for deriving an ordering is presented.

Ranking in Sports Tournaments

In many soccer leagues each team plays each other team twice. The winner of a match gets three points, in case of a tie both teams get one point. In the standard procedure, the final ranking of the teams in the championship is made up by adding these points and breaking ties by considering the goals scored. Another possible method of ranking the teams leads to a linear ordering problem. If there are n teams which have played each other twice we construct an (n, n)-matrix H by setting H_{ij} = number of goals which were scored by team i against team j. A triangulation of this matrix yields a ranking of the teams which takes the number of goals scored and (implicitly) the number of matches won into account. Moreover, transitive relations are important, and winning against a top team counts more than beating an average team.

As an example we compare the official ranking of the English Premier League in the season 2006/2007 with a ranking obtained by triangulation. Table 1.1 shows on the left side the official ranking and on the right side an optimum linear ordering. There are alternate optima, however, which we do not list here, but which would make this type of ranking approach problematical for practical use.

Table 1.1 Premier League 2006/2007 (left: official, right: triangulated)

1 Manchester United	1 Chelsea
2 Chelsea	2 Arsenal
3 Liverpool	3 Manchester United
4 Arsenal	4 Everton
5 Tottenham Hotspur	5 Portsmouth
6 Everton	6 Liverpool
7 Bolton Wanderers	7 Reading
8 Reading	8 Tottenham Hotspur
9 Portsmouth	9 Aston Villa
10 Blackburn Rovers	10 Blackburn Rovers
11 Aston Villa	11 Middlesborough
12 Middlesborough	12 Charlton Athletic
13 Newcastle United	13 Bolton Wanderers
14 Manchester City	14 Wigan Athletic
15 West Ham United	15 Manchester City
16 Fulham	16 Sheffield United
17 Wigan Athletic	17 Fulham
18 Sheffield United	18 Newcastle United
19 Charlton Athletic	19 Watford
20 Watford	20 West Ham United

Corruption Perception

The organisation *Transparency International* [160] releases an annual *corruption perception index* which ranks more than 150 countries by their perceived level of corruption. This index is computed from expert assessments and opinion surveys. The respective assessments and surveys only consider a subset of all countries and can thus also be viewed as a partial ordering. In [1] the linear ordering problem is used to aggregate these partial rankings. It is shown that the solution of the linear ordering problem agrees with the ranking according to the index to a large extent, but exhibits interesting differences for some countries.

Crossing Minimization

The relatively new research field *graph drawing* is concerned with finding good drawings of graphs for making the relations they represent easier to understand. In one of the many problems, the LOP could be employed. Let $G = (V, E)$ be a bipartite graph with the bipartition $V = V_1 \cup V_2$ of its nodes. A basic problem in graph drawing considers the task of drawing the nodes (linearly ordered) on two opposite horizontal lines and drawing the edges as straight lines such that the number of crossings is minimized. In [92] it is observed that the so-called *one sided crossing minimization problem* (where the permutation of one of the two sides is fixed) can be solved as a linear ordering problem with good results in practice. By embedding this procedure into a branch-and-bound algorithm, the two sided crossing minimization (without fixing) can also be solved. Figure 1.2 (taken from [92]) shows the results of a heuristic with 30 crossings and the optimal drawing with only 4 crossings.

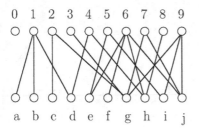 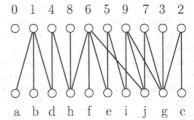

Fig. 1.2 Two sided crossing minimization

Linear Ordering with Quadratic Objective Function

Bipartite crossing minimization without fixed nodes can also be modelled directly employing the LOP. However in this case the objective function is not linear anymore, but contains products $x_{ij}x_{kl}$. A successful application of this model (using methods different from the ones presented in this book) is reported in [22].

Scheduling with Precedences

Consider a set of n jobs which have to be processed on a single machine. Each job i has a processing time p_i and a weight w_i. Furthermore, there is a set P of job pairs (i, j), each specifying that job i has to be executed before job j. (It is assumed that there are no contradictions within these precedences.) The task consists of finding a linear ordering $\langle k_1, k_2, \ldots, k_n \rangle$ of the jobs such that the total weighted completion time $\sum_{i=1}^{n} w_i t_i$ is minimized, where

$$t_{k_i} = \sum_{j=1}^{i} p_{k_j}$$

and the precedences given by P are observed. This problem can be modeled as a *linear ordering extension problem* where some relative rankings are already fixed beforehand. Solution methods are discussed in [16].

Linear Ordering with Cumulative Costs

Here, in addition to the arc weights c_{ij} of the standard linear ordering problem, there are node weights p_i, $1 \leq i \leq n$, and the task is to find an ordering $\langle k_1, k_2, \ldots, k_n \rangle$ of the nodes minimizing the cost $\sum_{i=1}^{j} \alpha_i$, where

$$\alpha_{k_i} = p_{k_i} + \sum_{j=i+1}^{n} c_{k_i k_j} \alpha_{k_j}.$$

An application of this problem for optimizing UMTS mobile phone telecommunication and its solution with a mixed-integer programming approach is discussed in [14]. Heuristic algorithms are presented in [50].

Coupled Task Problem

Many combinatorial optimization problems require as one constraint that some subset of elements is linearly ordered. We give a brief account to one of them where we could successfully use linear ordering variables in its optimization model [13]. The *coupled task problem* deals with scheduling n jobs each of which consists of two subtasks and where there is the additional requirement that between the execution of these subtasks an exact delay is required. If $\{J_1, J_2, \ldots, J_n\}$ is the set of jobs and $\{T_1, T_2, \ldots, T_{2n}\}$ the set of tasks, where T_{2i-1} and T_{2i} denote the first and second subtask of J_i, then one requirement for a feasible schedule is that all tasks are linearly ordered. We can model this constraint by introducing binary variables y_{kl} indicating whether task T_k is scheduled before task T_l or not. Of course, additional constraints are necessary to take processing times and gaps properly into account.

Target Visitation Problem

This optimization problem is a composition of the linear ordering problem and the traveling salesman problem and was proposed in [79]. Suppose that, starting from some origin 0, a set $\{1,\ldots,n\}$ of n targets has to be visited. In addition to the distance traveled, priorities have also to be taken into account. If d_{ij} (d_{0i} and d_{i0}) denotes the distance between two targets (the distance between the origin and a target and the distance between a target and the origin) and c_{ij} the gain when target i is visited before target j, the *target visitation problem* consists of finding a visiting sequence $\langle k_1, k_2, \ldots, k_n \rangle$ of the targets maximizing the objective function

$$\sum_{i=1}^{n-1} \sum_{j=i+1}^{n} c_{k_i k_j} - \left(d_{0k_1} + \sum_{i=1}^{n-1} d_{k_i k_{i+1}} + d_{k_n 0} \right).$$

In the literature, the term "linear ordering problem" is sometimes misused also for some problems where a linear ordering of the nodes of some weighted graph has to be found, but where the objective function is a different one. For example, [3] actually considers the *linear arrangement problem* and [95] requests as the objective that the capacity of a cut between two adjacent nodes is as small as possible, i.e., it deals with the *cut-width problem*.

1.1.2 The LOLIB Library of Benchmark Instances

We have compiled a set of benchmark problem instances of the LOP. These problem instances consist of real-world as well as randomly generated data sets.

Data Format

For finding the optimum triangulation of a matrix, the diagonal entries are irrelevant. Furthermore, orderings compare in the same way if some constant C is added to both entries H_{ij} and H_{ji}. In particular, the optimality of an ordering is not affected by this transformation. However, the quality of bounds does change. If we take diagonal entries into account and add a large constant to every matrix entry, then every feasible solution is close to optimal and no real comparison of qualities is possible. Therefore we transform every problem instance to a suitable normal form.

Definition 1.1. A quadratic (n,n)-matrix H is in *normal form* if

   (i) all entries of H are integral and nonnegative,
   (ii) $H_{ii} = 0$, $i = 1,\ldots,n$,
   (iii) $\min\{H_{ij}, H_{ji}\} = 0$, $1 \leq i < j \leq n$.

The following example shows a matrix and its normal form.

$$\begin{pmatrix} -17 & 36 & 11 & 45 & 7 \\ 21 & 22 & 44 & 15 & 9 \\ 26 & 23 & 13 & 26 & 12 \\ 22 & 22 & 11 & 0 & 33 \\ 30 & 9 & 25 & 24 & -7 \end{pmatrix} \quad \begin{pmatrix} 0 & 15 & 0 & 23 & 0 \\ 0 & 0 & 21 & 0 & 0 \\ 15 & 0 & 0 & 15 & 0 \\ 0 & 7 & 0 & 0 & 9 \\ 23 & 0 & 13 & 0 & 0 \end{pmatrix}$$

For our computations, all matrices H are transformed to their normal form $\bar{H}$. Note that this normal form is unique. Since the "true" value of a linear ordering might be interesting for some application, we compute

$$s_d = \sum_{i=1}^{n} H_{ii} \text{ and } s_t = \sum_{i-1}^{n-1} \sum_{j-i+1}^{n} \min\{H_{ij}, H_{ji}\}.$$

Now, if $c_{\text{opt}}(\bar{H})$ is the value of an optimum triangulation of $\bar{H}$, then $c_{\text{opt}}(H) = c_{\text{opt}}(\bar{H}) + s_d + s_t$.

In normal form, a matrix can be seen as the specification of a weighted tournament (take all arcs (i, j) with $H_{ij} > 0$ and, if $H_{ij} = H_{ji} = 0$, then choose one of the arcs (i, j) or (j, i)). The LOP then directly corresponds to finding a minimum weight feedback arc set for this tournament.

Definition 1.2. Let H be a matrix and σ be an optimum linear ordering. Then the number

$$\lambda(H) = \frac{\sum\limits_{\sigma(i)<\sigma(j)} H_{ij}}{\sum\limits_{i \neq j} H_{ij}}$$

is called *degree of linearity* of H.

The degree of linearity gives the sum of the superdiagonal entries as the percentage of the total sum of the matrix entries (except for the diagonal elements) and allows for some interpretations in economical analysis. It lies between 0.5 (this is a trivial lower bound) and 1.0 (for a triangular matrix) and is an indicator for the closeness of a matrix to a triangular matrix. Note, that the degree of linearity differs depending on whether a matrix is in normal form or not, so some care has to be given when interpreting it.

From a computational point of view it turns out that problems with smaller degree of linearity tend to become more difficult. This is validated by some experiments with random matrices.

We now describe the problem instances selected for the benchmark library.

Input-Output Matrices

These are real-world data sets taken from input-output tables from various sources. The corresponding linear ordering problems are comparatively easy. They are thus more of interest for economists than for the assessment of heuristics for hard problems. So we have not conducted extensive experiments with them. Only the quality of simple heuristics can be assessed with these matrices. The original entries in these tables were not necessarily integral, but for LOLIB they were scaled to integral values.

European 44-Sector Tables

The Statistical Office of the European Communities (Eurostat) compiles input-output tables for the member states of the EC. Our benchmark set contains 31 matrices of dimension 44 from the years 1959 to 1975 (t59b11xx – t75u11xx).

Belgian 50-Sector Tables

These input-output tables (be75eec, be75np, be75oi and be75tot) of 1975 with 50 sectors were compiled for the Belgian economy.

German 56-Sector Tables

These matrices (tiw56n54 – tiw56r72) for some years between 1954 to 1975 were compiled by *Deutsches Institut für Wirtschaftsforschung (DIW)* for the *Federal Republic of Germany*.

German 60-Sector Tables

These input-output tables (stabu70, stabu74 and stabu75) were compiled by the *Statistisches Bundesamt* of the *Federal Republic of Germany* for the years 1970, 1974 and 1975. (In some publications, these matrices were named stabu1 – stabu3.)

US 79-Sector Table

The matrix usa79 is the input-output table for the economy of the United States for the year 1985. It has been made available by Knuth [99].

Randomly Generated Instances A (Type 1)

This is a set of random problems defined by Martí [117] that has been widely used for experiments. Problems are generated from a (0,100) uniform distribution. Sizes are 100, 150, 200 and 500 and there are 25 instances in each set for a total of 100. The names of these instances are t1dn.i (e.g. t1d200.25), where n is the dimension and i the number within the instances of the same size.

In their original definition, these problem instances are not in normal form. For the experiments in this monograph and for publication in the library LOLIB we only give values with respect to the normalized objective function.

Randomly Generated Instances A (Type 2)

This data set has also been defined by Martí [117]. Problems are generated by counting the number of times a sector appears in a higher position than another in a set of randomly generated permutations. For a problem of size n, $n/2$ permutations are generated.

There are 25 instances with sizes 100, 150 and 200, respectively. The names of these instances are t2dn.i (e.g. t2d150.12), where n is the dimension and i the number within instances of the same size.

Randomly Generated Instances B

In this kind of random problem we tried to influence the difficulty of the problems for computational experiments. To this end we generated integer matrices where the superdiagonal entries are drawn uniformly from the interval $[0, U_1]$ and the subdiagonal entries from $[0, U_2]$, where $U_1 \geq U_2$. The difference $U_1 - U_2$ affects the difficulty. Subsequently, the instances were transformed to normal form and a random permutation was applied.

For $n = 40$, we set $U_1 = 100$ and $U_2 = 100 + 4(i - 1)$ for problems p40-i. For $n = 44$ and $n = 50$ we set $U_1 = 100$ and $U_2 = 100 + 2(i - 1)$ for problems p44-i, p50-i, respectively.

SGB Instances

These instances were used in [106] and are taken from the *Stanford GraphBase* [99]. They are random instances with entries drawn uniformly distributed from $[0, 25000]$. The set has a total of 25 instances with $n = 75$. Instances are named sgb75.01 through sgb75.25.

Instances of Schiavinotto and Stützle

Some further benchmark instances have been created and used by Schiavinotto and Stützle [154]. These instances were generated from the input-output tables by replicating them to obtain larger problems. Thus, the distribution of numbers in these instances somehow reflects real input-output tables, but otherwise they behave more like random problems. This data set has been called XLOLIB, and instances with $n = 150$ and $n = 250$ are available.

Instances of Mitchell and Borchers

These instances have been used by Mitchell and Borchers for their computational experiments [127]. They are random matrices where the subdiagonal entries are uniformly distributed in $[0, 99]$ and the superdiagonal entries are drawn uniformly from $[0, 39]$. Furthermore a certain percentage of the entries was zeroed out.

Further Special Instances

We added some further instances that were used in some publications.

EX Instances

These random problems (EX1–EX6) were used in particular in [39] and [40].

econ Instances

The problem instances econ36 through econ77 were generated from the matrix usa79. They turned out not to be solvable as a linear program using only 3-dicycle inequalities.

Paley Graphs

Paley graphs, or more precisely *Paley tournaments*, have been used by Goemans and Hall [75] to prove results about the acyclic subdigraph polytope. These tournaments are defined as follows. Let $q = 3 \mod 4$ be a prime power. Define the digraph on q nodes corresponding to the elements of the finite field $GF(q)$. This digraph contains the arc (i, j) if and only if $j - i$ is a nonzero square in $GF(q)$. Some of them provide interesting difficult linear ordering problems.

atp Instances

These instances were created from the results of the ATP tennis tour 1993/1994. Nodes correspond to a selection of players and the weight of an arc (i, j) is the number of victories of player i against player j.

Table 1.2 summarizes the number of instances in each set described above. Moreover, it specifies the number of instances for which either the optimum or only an upper bound is known. In the computational experiments we call the set of 229 instances for which the optimum is known OPT-I, and the set of 255 instances for which an upper bound is known UB-I.

Table 1.2 Number of instances in each set

Set	#Instances	#Optima	#Upper Bounds
IO	50	50	–
SGB	25	25	–
RandomAI	100	–	100
RandomAII	75	25	50
RandomB	90	70	20
MB	30	30	–
XLOLIB	78	–	78
Special	36	29	6
Total	484	229	255

LOLIB is available at the web site www.uv.es/rmarti/paper/lop.html and it is referenced as [117]. Also the currently best known values and upper bounds as well as the constants eliminated by the transformation to normal form can be found there.

1.2 The Maximum Diversity Problem

The problem of maximizing diversity deals with selecting a subset of elements from a given set in such a way that the diversity among the elements is maximized. It was first approached from an Operations Research perspective in 1988 by Kuhy [54], and presented in 1993 in the annual meeting of the Decision Science Institute, where Kuo, Glover, and Dhir, proposed integer programming models [104], and applied them to preserving biological diversity in [67]. Several models have been proposed to deal with this combinatorial optimization problem since then. All of them require a diversity measure, typically a distance function in the space where the objects belong, which is customized to each specific application. Diversity models have been applied to product design, social problems or workforce management [68] to mention a few.

In its graph version the MDP is defined as follows. Let $G_n = (V_n, E_n)$ denote the complete graph on n nodes, i.e., the undirected graph with node set $V_n = \{1, 2, \dots, n\}$ where for every pair of nodes i and j there is an edge (i, j) joining them. The edge (i, j) can be also denoted as (j, i). Each edge (i, j) has an associated weight or distance, d_{ij}. A feasible solution of the MDP is a set $M \subset V_n$ of m elements, and its value is the sum of distances between the pairs of nodes in M.

> **Maximum diversity problem**
> Given the complete graph $G_n = (V_n, E_n)$ with edge distances d_{ij} for every pair $i, j \in V_n$ and an integer m, compute a subset M of V_n such that $|M| = m$ and $\sum_{i,j \in M} d_{ij}$ is as large as possible.

When formulating a model for a specific application, we compute the distance from problem data. It is assumed that each element in the application can be represented by a set of K attributes. Let s_{ik} be the state or value of the $k - th$ attribute of element i, where $k = 1, \cdots, K$. Then, the distance between elements i and j may be computed as

$$d_{ij} = \sqrt{\sum_{k=1}^{K} \left(s_{ik} - s_{jk} \right)^2}$$

In this case, d_{ij} is the Euclidean distance computed from the attributes of i and j, and in location-based problems, it is simply the distance from the points coordinates. However, other distance functions can be considered as well, even those taking negative values. Sandoya et al. [151] described the cosine similarity, an affinity measure in the context of a model to maximize the equity, computed as:

$$d_{ij} = \frac{\sum_{k=1}^{K} s_{ik} s_{jk}}{\sqrt{\sum_{k=1}^{K} s_{ik}^2} \sqrt{\sum_{k=1}^{K} s_{jk}^2}}.$$

The cosine similarity between two elements can be viewed as the angle between their attribute vectors, where a small angle between elements indicates a large degree of similarity. It takes values in $[-1, 1]$ reflecting the affinity between the individuals that may be positive or negative.

The MDP can be trivially formulated in mathematical terms as a quadratic binary problem, where variable x_i takes the value 1 if element i is selected and 0 otherwise, $i = 1, \ldots, n$.

$$\max z_{MS}(x) = \sum_{i<j} d_{ij} x_i x_j$$

$$\sum_{i=1}^{n} x_i = m$$

$$x_i \in \{0,1\}, i = 1, \ldots, n.$$

Kuo, Glover and Dhir [67] use this formulation to show that the clique problem (which is known to be NP-complete) is reducible to the MDP.

The following classic example, described in [52], illustrates that the term diversity is somehow ambiguous in the context of combinatorial optimization, and some problems seem to look for dispersion among the selected points, while others want to achieve some representativeness, where the selected points are representing a class in the given set.

"Consider a group of university students from which we want to select five to form a committee. For each pair of students, we can compute a distance value based on their particular attributes. These include personal characteristics such as age or gender. A pair of students with similar attributes receives a low distance value, while students with different attributes have a large distance value. To create a diverse committee, the five students have to be selected in a way that their ten pairwise distances are relatively large. The most diverse committee would represent the maximum sum of the distances between the selected students.".

It is clear that in the example above, by a *diverse committee* we refer to a set of students that represents well the entire group in that university. In other words, it seems that the objective is that the selected students contain most of the attributes in the group, rather than their dispersion in the group. We can classify it as an instance of *representative models*. In line with this, Page states in his book, *The difference – How the power of diversity creates better groups, firms, schools, and societies* [136], that diverse perspectives and tools enable collections of people to find more and better solutions and contribute to overall productivity. As a result, the problem of identifying diverse groups of people becomes a key point in large firms and institutions.

Other types of applications fall into the class of *equity models*. They are mainly used in the context of facility location problems, where the fairness among candidate facility locations is as relevant as the dispersion of the selected locations [53]. These kinds of problems have their applications in the context of urban public location.

The third group of applications, and probably the most important one, is the direct implementation of the MDP. Many applications, such as telecommunications or transportation, look for diverse set of items, such as routes or nodes, that are disperse (in many settings it refers to geographically disperse). One of the early applications of *dispersion models* deals with preserving biological diversity [67].

Biological diversity refers to the richness and equal abundance of species. The study of biodiversity can be traced back to the eighties, when early investigators of the natural world provided a succinct summary of the relevant work on ecological diversity, including a practical guide to its measurement. Maximizing biological diversity first requires selecting a measure of dissimilarity. The study in [67] relies on the operational taxonomic units (OTUs) to compute a dissimilarity between two OTUs with respect to some attributes of interest, such as petal color, unison call, or number of gene substitution per locus. From such a quantitative definition the authors applied the first mathematical programming model, the MDP, to compute and maximize the diversity in this context.

We now describe in mathematical terms the different models proposed in the context of dispersion, equity, and representativeness.

1.2.1 Diversity measures and models

As mentioned above, the MDP is a classic model to capture the notion of diversity. However, researchers have proposed over the years alternative models to reflect diversity, dispersion or even equity.

Max-Min Diversity Problem

Early papers from 1993 already proposed an interesting model that maximizes the minimum distance between the selected elements [104]. It is called the *Max-Min Diversity Problem* (MMDP), and can be formulated in similar terms than the MDP as follows.

$$\max \, z_{MM}(x) = \min_{i,j \in M} d_{ij} x_i x_j$$

$$\sum_{i=1}^{n} x_i = m$$

$$x_i \in \{0,1\}, \, i = 1, \dots, n.$$

The set of solutions M in the formulation above can be represented as $M = \{i \in V_n : x_i = 1\}$. Although the MDP and the MMDP are related, Resende et al. [146] illustrate with the following example that the correlation between the values of the solutions in both problems can be relatively low.

$$\begin{pmatrix} 0.0 & 4.6 & 6.2 & 2.1 & 3.5 & 3.6 & 4.4 \\ 4.6 & 0.0 & 6.6 & 7.1 & 8.2 & 2.4 & 5.3 \\ 6.2 & 6.6 & 0.0 & 2.1 & 3.5 & 3.6 & 4.4 \\ 2.1 & 7.1 & 2.1 & 0.0 & 5.5 & 1.1 & 2.3 \\ 3.5 & 8.2 & 3.5 & 5.5 & 0.0 & 6.4 & 3.4 \\ 3.6 & 2.4 & 3.6 & 1.1 & 6.4 & 0.0 & 5.4 \\ 4.4 & 5.3 & 4.4 & 2.3 & 3.4 & 5.4 & 0.0 \end{pmatrix}$$

The matrix above contains the distances between the pairs of seven elements of which we need to select five. For such a small example, we can enumerate all possible solutions (selections of $m = 5$ out of $n = 7$ elements) and compute for each its MDP and MMDP values ($z_{MS}(x)$ and $z_{MM}(x)$ respectively). The correlation between both objective functions is 0.52, which can be considered relatively low. Moreover, we find that the optimal solution x^* of the MDP has a value $z_{MS}(x^*) = 54.4$ and a value $z_{MM}(x^*) = 2.1$. However, the optimal solution y^* of the MMDP has a value $z_{MM}(y^*) = 3.3$, which is relatively larger than $z_{MM}(x^*)$. Moreover, 30% of the solutions present a $z_{MM}(x)$ value larger than $z_{MM}(x^*)$. Resende et al. [146] conclude their analysis by saying that we should not expect a method for the MDP to obtain good solutions for the MMDP or viceversa, and therefore we need to design specific solving methods for each model.

As will be shown, the Max-Min problem is an instance of both dispersion and representative models, since it looks for a well-distributed set of points in the solution space.

Equity models

Prokopyev et al. [139] introduced two other models, which incorporate the concept of fairness among candidates. These models appear in different problems, such as facility location or group selection, in which we look for fair diversification. The *Maximum MinSum dispersion problem* maximizes the minimum aggregate dispersion among the chosen elements, while the *Minimum Differential dispersion problem* minimizes extreme equity values of the selected elements.

The Maximum MinSum dispersion problem (Max-MinSum) consists of selecting a set $M \subseteq V$ of m elements such that the smallest total dispersion associated with each selected element i is maximized. The problem is formulated in [139] as follows:

$$\max \left\{ \min_{i:x_i=1} \sum_{j:j \neq i} d_{ij} x_j \right\}$$

$$\sum_{i=1}^{n} x_i = m$$

$$x_i \in \{0,1\}, \ i = 1, \ldots, n.$$

The Minimum Differential dispersion problem (Min-Diff) consists in finding the best subset $M \subseteq V$ with respect to the difference between the maximum and the minimum of the sum of distances among the selected elements. This problem is a 0-1 integer linear programming problem and can be formulated as:

$$\min \left\{ \max_{i:x_i=1} \sum_{j:j \neq i} d_{ij} x_j - \min_{i:x_i=1} \sum_{j:j \neq i} d_{ij} x_j \right\}$$

$$\sum_{i=1}^{n} x_i = m$$

$$x_i \in \{0,1\}, \ i = 1, \ldots, n.$$

Generalized models

Prokopyev et al. [139] also proposed two additional models that generalize the previous ones.

In the *Max Mean dispersion problem* (Max-Mean), the typical cardinality restriction of the diversity and dispersion models is not imposed. This problem is a version of the MDP where the number of elements to be selected is unknown. It can be formulated as the following 0-1 integer linear programming problem:

$$\text{Maximize} \quad \frac{\sum_{i=1}^{n-1} \sum_{j=i+1}^{n} d_{ij} x_i x_j}{\sum_{i=1}^{n} x_i} \tag{1.1}$$

$$\text{subject to: } \sum_{i=1}^{n} x_i \geq 2$$
$$x_i \in \{0,1\}, \ i = 1, \dots, n.$$

A more general version of this problem, which includes weights on the elements, is called *Generalized Max-Mean Dispersion Problem* and is formulated as follows

$$\text{Maximize} \quad \frac{\sum_{i=1}^{n-1} \sum_{j=i+1}^{n} d_{ij} x_i x_j}{\sum_{i=1}^{n} w_i x_i} \tag{1.2}$$

$$\text{subject to: } \sum_{i=1}^{n} x_i \geq 2$$
$$x_i \in \{0,1\}, \ i = 1, \dots, n.$$

where w_i is the weight assigned to element $i \in V$. Table 1.3 summarizes the six diversity measures described in the models above.

Table 1.3 Diversity measures.

Measure	Mathematical function	Model
Sum	$\sum_{i<j, i,j \in M} d_{ij}$	MDP or Max-Sum
Min	$\min_{i<j, i,j \in M} d_{ij}$	MMDP or Max-Min
Minsum	$\min_{i \in M} \sum_{j \in n, j \neq i} d_{ij}$	MaxMinSum
Difference	$\max_{i \in M} \sum_{j \in M, j \neq i} d_{ij} - \min_{i \in M} \sum_{j \in M, j \neq i} d_{ij}$	MinDiff
Mean	$\dfrac{\sum_{i<j, i,j \in M} d_{ij}}{|M|}$	MeanDP
Mean	$\dfrac{\sum_{i<j, i,j \in M} d_{ij}}{\sum_{i=1}^{n} w_i x_i}$	Generalized MeanDP

Comparison of models

Parreño et al. [138] perform an empirical comparison of four models: MaxMin, MaxSum, MaxMinSum, and MinDiff. In particular, the authors analyze these models from both numerical and geometrical perspectives, comparing the structure of their respective solutions obtained with CPLEX.

The first conclusion in [138] is that the MaxSum and MaxMinSum provide similar solutions, and considering the relatively large amount of research already done in the MaxSum model, it is not well justified the need of the recently introduced MaxMinSum. Figure 1.3 compares the structure of the solutions obtained with both models on an instance with $n = 50$ points where we want to select $m = 10$ of them.

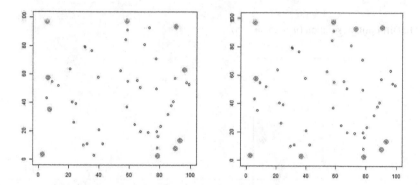

Fig. 1.3 MaxMinSum (left) and MaxSum (right) optimal solutions ($n = 50$ and $m = 10$)

The geometrical configuration of the selected points in Figure 1.3 shows that the MaxSum and MaxMinSum models obtain solutions close to the borders, and with no points in the central region. This clearly induces diversity or dispersion in the set of selected points, but the authors point out that the models do not select any solution in the central region, which can be interpreted as a lack of representativeness.

The second conclusion in [138] is that the MinDiff model only seeks for inter-distance equality among the selected points, and ignores how large or small these distances are. Figure 1.4 shows the optimal MinDiff solution of an instance with $n = 25$ and $m = 5$, where the solution seeks for 5 elements with equal or at least very similar inter-distance values, which as we can see, are obtained in the corner points of a regular pentagon. The authors recommend avoid the use of this model to achieve diversity or dispersion, unless very well justified for a specific application.

Finally, Parreño et al. [138] conclude that the MaxMin model generates solutions with a very different structure than the MaxSum model. It obtains equidistant points all over the space, and it does not avoid to select points in the central part, which can be interpreted as representativeness, more than dispersion. Figure 1.5 shows the optimal MaxMin solution of an instance with $n = 50$ and $m = 10$, which illustrates this point.

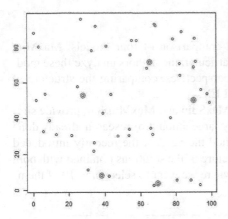

Fig. 1.4 MinDiff optimal solution ($n = 25$, $m = 5$)

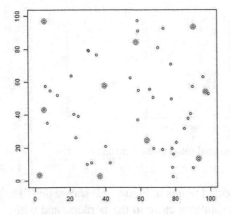

Fig. 1.5 MaxMin optimal solution ($n = 50$, $m = 10$)

From the analysis above, we consider the MaxSum model in our study since it truly achieves diversity. Additionally, it has been extensively studied, with many optimization methods proposed to solve it, which makes the MDP especially interesting to describe a wide range of optimization technologies.

1.2.2 The MDPLIB Library of Benchmark Instances

Martí and Duarte [116] compiled in 2010 a comprehensive set of benchmark instances representative of the collections used for computational experiments in the MDP. They called this benchmark MDPLIB , and collects a total of 315 instances, available at www.uv.es/rmarti/paper/mdp.html.

In the MDPLIB all the instances have been randomly generated. The generators were not built according to any specific application or to mimic the type of instances found in a real example. On the contrary, they are rather artificial, created with the purpose of being a challenge for heuristic methods. This is one of the main differences with the LOLIB in which, as shown in the previous section, we can find real instances and random instances created in line with the real ones.

The MDPLIB basically contains three sets of instances collected from different papers: GKD, MDG, and SOM. We now describe in detail each set of instances, which contains different subsets according to their source. We also reference the authors who generated them.

GKD

This data set consists of 145 matrices for which the values were calculated as the Euclidean distances from randomly generated points with coordinates in the 0 to 10 range. It collects three subsets, namely GKD-a, GKD-d, and GKD-c.

GKD-a: Glover et al. [68] introduced these 75 instances in which the number of coordinates for each point is generated randomly in the 2 to 21 range, and the distance values are computed with the Euclidean formula from these coordinates. The instance sizes are such that for $n = 10$, $m = 2, 3, 4, 6$ and 8; for $n = 15$, $m = 3, 4, 6, 9$ and 12; and for $n = 30$, $m = 6, 9, 12, 18$ and 24.

GKD-b: Martí et al. [115] generated these 50 matrices for which the number of coordinates for each point is generated randomly in the 2 to 21 range and the instance sizes are such that for $n = 25$, $m = 2$ and 7; for $n = 50$, $m = 5$ and 15; for $n = 100$, $m = 10$ and 30; for $n = 125$, $m \doteq 12$ and 37; and for $n = 150$, $m = 15$ and 45.

GKD-c: Duarte and Martí [51] generated these 20 matrices with 10 coordinates for each point and n = 500 and m = 50. The distance values are computed, as in the other GKD-subsets, with the Euclidean formula from these coordinates.

MDG

This data set consists of 100 matrices with real numbers randomly selected according to a uniform distribution.

MDG-a. Duarte and Martí [51] generated these 40 matrices with real numbers randomly selected in [0, 10], 20 of them with $n = 500$ and $m = 50$ and the other 20 with $n = 2000$ and $m = 200$. These instances were used in [137].

MDG-b. This data set consists of 40 matrices generated with real numbers randomly selected in [0, 1000] [51]. 20 of them have $n = 500$ and $m = 50$, and the other 20 have $n = 2000$ and $m = 200$. These instances were used in [61] and [137].

MDG-c. Considering that many heuristics were able to match the best-known results in many of the instances previously introduced, Martí et al. [116] proposed this data set with very large instances in 2013. It consists of 20 matrices with randomly generated numbers according to a uniform distribution in the range [0, 1000], and with $n = 3000$ and $m = 300, 400, 500$ and 600. These are the largest instances reported in our computational study.

SOM

This data set consists of 70 matrices with integer random numbers between 0 and 9 generated from an integer uniform distribution. It is split into two subsets, SOM-a and SOM-b.

SOM-a. These 50 instances were generated by Martí et al. [115] with a generator developed by Silva et al. [153] in 2004. The instance sizes are such that for $n = 25$, $m = 2$ and 7; for $n = 50$, $m = 5$ and 15; for $n = 100$, $m = 10$ and 30; for $n = 125$, $m = 12$ and 37; and for $n = 150$, $m = 15$ and 45.

SOM-b. These 20 instances were generated by Silva et al. [153] and used in most of the previous papers (see for example Aringhieri et al. [8]. The instance sizes are such that for $n = 100$, $m = 10, 20, 30$ and 40; for $n = 200$, $m = 20, 40, 60$ and 80; for $n = 300$, $m = 30, 60, 90$ and 120; for $n = 400$, $m = 40, 80, 120$, and 160; and for $n = 500$, $m = 50, 100, 150$ and 200.

We end the description of the instances, summarizing them in Table 1.4. This table shows the number of instances and the range of n and m in each subset.

Table 1.4 Number of instances in the MDPLIB

Set	#Instances	Range of n	Range of m
GKD-a	75	[10, 30]	[2, 24]
GKD-b	50	[25, 150]	[2, 45]
GKD-c	20	500	50
MDG-a	40	[500, 2000]	[50, 200]
MDG-b	40	[500, 2000]	[50, 200]
MDG-c	20	3000	[300, 600]
SOM-a	50	[25, 150]	[2, 45]
SOM-b	20	[100, 500]	[10, 200]
Total	315	[10, 3000]	[2, 600]

Martí et al. [114] published this year 2021 an updated version of the library, called MDPLIB 2.0, in which some other instances used in diversity problems are included. In particular the new sets are:

GKD-d. Parreño et al. [138] generated 70 matrices for which the values were calculated as the Euclidean distances from randomly generated points with two co-ordinates in the 0 to 100 range. For each value of $n = 25, 50, 100, 250, 500, 1000$, and 2000, they considered 10 instances with $m = \lceil n/10 \rceil$ and 10 instances with $m = 2\lceil n/10 \rceil$, totalizing 140 instances. The main motivation of this new set is to include the original coordinates in the instances files that unfortunately are not publicly available nowadays for the other subsets. In this way, researchers may represent the solutions in line with the work in [138].

ORLIB. This is a set of 10 instances with $n = 2500$ and $m = 1000$ that were proposed for binary problems [11]. The distances are integers generated at random in $[-100, 100]$ where the diagonal distances are ignored.

PI. Palubeckis [137] generated 10 instances where the distances are integers from a $[0, 100]$ uniform distribution. 5 of them are generated with $n = 3000$ and $m = 0.5n$, and 5 with $n = 5000$ and $m = 0.5n$. The density of the distance matrix is 10%, 30%, 50%, 80% and 100%.

MGPO. To complement the sets above, 80 large matrices with relatively low m values were generated. They are 40 instances with $n = 1000$ and integer numbers randomly selected in $[1, 100]$, 20 of them with $m = 50$ and 20 with $m = 100$, and 40 matrices with $n = 2000$ and integer numbers randomly selected in $[1, 100]$, 20 of them with $m = 50$, and 20 with $m = 100$.

In the MDPLIB 2.0 [114] the authors replaced the old sets, GKD-a and GKD-b, containing small instances with the four sets above with larger and more challenging instances. Additionally, they included 300 instances for constrained versions of diversity problems that include cost and capacity values. MDPLIB 2.0 has 770 instances and is available at www.uv.es/rmarti/paper/mdp.html.

Chapter 2
Heuristic Methods

Abstract Since the linear ordering problem and the maximum diversity problem
are both NP-hard, we cannot expect to be able to solve practical problem instances
of arbitrary size to optimality. Depending on the size of an instance or depending on
the available CPU time we will often have to be satisfied with computing approximate solutions. In addition, under such circumstances, it might be impossible to
assess the real quality of approximate solutions. In this and in the following chapter
we will deal with the question of how to find very good solutions for the LOP and
the MDP in short or reasonable time.
The methods described in this chapter are called *heuristic algorithms* or simply
heuristics. This term stems from the Greek word *heuriskein* which means to find
or discover. It is used in the field of optimization to characterize a certain kind of
problem-solving methods. There are a great number and variety of difficult problems, which come up in practice and need to be solved efficiently, and this has pro-
moted the development of efficient procedures in an attempt to find good solutions,
even if they are not optimal. These methods, in which the process speed is as important as the quality of the solution obtained, are called heuristics or *approximate*
algorithms.

2.1 Introduction

As opposed to *exact methods*, which guarantee to provide an optimum solution of
the problem, heuristic methods only attempt to yield a good, but not necessarily
optimal solution. Nevertheless, the time taken by an exact method to find an optimum solution to a difficult problem, if indeed such a method exists, is in a much
greater order of magnitude than the heuristic one (sometimes taking so long that in
many cases it is inapplicable). Thus we often resort to heuristic methods to solve
real optimization problems.

Perhaps the following comment by Onwubolu and Babu [133] is a little far-
fetched: "The days when researchers emphasized using deterministic search tech-

© Springer-Verlag GmbH Germany, part of Springer Nature 2022
R. Martí and G. Reinelt, *Exact and Heuristic Methods in
Combinatorial Optimization*, Applied Mathematical Sciences 175,
https://doi.org/10.1007/978-3-662-64877-3_2

niques to find optimal solutions are gone.". But it is true that in practice an engineer, an analyst or a manager sometimes might have to make a decision as soon as possible in order to achieve desirable results.

Recent years have witnessed a spectacular growth in the development of heuristic procedures to solve optimization problems. This fact is clearly reflected in the large number of articles published in specialized journals. 1995 saw the first issue of the *Journal of Heuristics*, dedicated solely to the publication of heuristic procedures. In the same year the first international congress dealing with these methods, called the *Metaheuristic International Conference (MIC)*, was held in Breckenridge, Colorado (USA). Since then, this conference has taken place every other year with more and more participants; thus consolidating what is known now as the metaheuristics community, with more than 1,400 researchers from over 80 countries, and presence in Europe as a EURO working group (EU/ME- https://www.euro-online.org/websites/eume/).

In addition to the need to find good solutions of difficult problems in reasonable time, there are other reasons for using heuristic methods, among which we want to highlight:

–   No method for solving the problem to optimality is known.
–   Although there is an exact method to solve the problem, it cannot be used on the available hardware.
–   The heuristic method is more flexible than the exact method, allowing, for example, the incorporation of conditions that are difficult to model.
–   The heuristic method is used as part of a global procedure that guarantees to find the optimum solution of a problem.

A good heuristic algorithm should fulfil the following properties:

–   A solution can be obtained with reasonable computational effort.
–   The solution should be near optimal (with high probability).
–   The likelihood for obtaining a bad solution (far from optimal) should be low.

There are many heuristic methods that are very different in nature. Therefore, it is difficult to supply a full classification. Furthermore, many of them have been designed to solve a specific problem without the possibility of generalization or application to other similar problems. The following outline attempts to give wide, non-excluding categories, under which to place the better-known heuristics:

Decomposition Methods

The original problem is broken down into sub-problems that are simpler to solve, bearing in mind, be it in a general way, that subproblems belong to the same problem class.

Inductive Methods

The idea behind these methods is to generalize the smaller or simpler versions to the whole case. Properties or techniques that have been identified in these cases which are easier to analyze, can be applied to the whole problem.

Reduction Methods

These involve identifying properties that are mainly fulfilled by the good solutions and introduce them as boundaries to the problem. The objective is to restrict the space of the solutions by simplifying the problem. The obvious risk is that the optimum solutions of the original problem may be left out.

Constructive Methods

These involve building a solution to the problem literally step by step from scratch. Usually they are deterministic methods and tend to be based on the best choice in each iteration. These methods have been widely used in classic combinatorial optimization.

Local Search Methods

In contrast to the methods previously mentioned, local improvement or local search starts with some feasible solution of the problem and tries to progressively improve it. Each step of the procedure carries out a movement from one solution to another one with a better value. The method terminates when, for a solution, there is no other accessible solution that improves it.

Even though all these methods have contributed to expanding our knowledge of solving real problems, the constructive and local search methods form the foundations of the meta-heuristic procedures [4], which will be described in the next chapter.

2.1.1 Assessing the Quality of Heuristics

There are diverse possibilities for measuring the quality of a heuristic, among which we find the following.

Comparison with the Optimum Solution

Although one normally resorts to an approximative algorithm, because no exact method exists to obtain an optimum solution or it is too time-consuming, sometimes a procedure is available that provides an optimum for a limited set of examples (usually small sized instances). This set of examples can be used to assess the quality of the heuristic method.

Normally, for each example, the following are measured: the percentaged deviation of the heuristic solution value as compared to the optimum one and the mean of these deviations. If we denote by c_A the value of the solution delivered by heuristic A and by c_{opt} the optimum value of a given example, in a maximization problem like the LOP, the percentaged deviation, *PerDev*, is given by the expression

$$PerDev = 100 \cdot \frac{c_{opt} - c_A}{c_{opt}}.$$

(We assume that all feasible solutions have a positive value.)

Comparison with a Bound

There are situations when no optimum solution is available for a problem, not even for a limited set of examples. An alternative evaluation method involves comparing the value of the solution provided by the heuristic with a bound for the problem (a lower bound if it is a minimization problem and an upper bound if it is a maximization problem). Obviously the quality of fit will depend on the quality of the bound (closeness to optimal). Thus we must somehow have information about the quality of the aforementioned bound, otherwise the proposed comparison would not be of much interest.

Comparison with a Truncated Exact Method

An enumerative method like branch-and-bound explores very many solutions, even though this may be a fraction of the total, and therefore large-scale problems can be computationally out of reach using these methods. Nevertheless, we can establish a limit on the maximum number of iterations (or on the CPU time) to run the exact algorithm. Moreover, we can modify the criteria to fathom a node in the search tree by adding or subtracting (depending on whether it is a minimization or maximization problem) a value Δ to the bound of the node thus fathoming a larger number of nodes and speeding up the method. In this way it guarantees that the value of the best solution provided by the procedure is no further than distance Δ from the optimal value to the problem. In any case, the best solution found with these truncated procedures establishes a bound against which the heuristic can be measured.

Comparison with Other Heuristics

This is one of the most commonly used methods for difficult problems which have been worked on for a long time and for which some good heuristics are known. Similarly to what happens with the bound comparisons, the conclusion of this comparison deals with the quality of fit of the chosen heuristic.

Given that the LOP has been studied in-depth from both the exact viewpoint and that of a heuristic, we have a value of the optimum solution for small and medium-scale examples, which enables us to establish the optimal deviation in the solution obtained by the heuristics. Furthermore, we can compare the values obtained between the different heuristics to solve the same examples of any size.

Worst Case Analysis

One method that was well-accepted for a time concerns the behavioral analysis of the heuristic algorithm in the worst case; i.e., consider the examples that most disfavor the algorithm and set analytical bounds to the maximal deviation in terms of the optimum solution to the problem. The best aspect of this method is that it established the limits of the algorithm's results for any example. However, for the same reason, the results tend not to be representative of the average behavior of the algorithm. Furthermore, the analysis can be very complicated for more sophisticated heuristics.

An algorithm A for dealing with a maximization problem is called *ε-approximative* if there is a constant $\varepsilon > 0$ such that for every problem instance the algorithm guarantees that a feasible solution can be found with value c_A and the property

$$c_A \geq (1 - \varepsilon)c_{\text{opt}}.$$

The analogous definition for minimization problems is $c_A \leq (1 + \varepsilon)c_{\text{opt}}$.

Concerning the approximability of the LOP the following results are known. Suppose that all objective function coefficients are nonnegative and take some arbitrary ordering. Then either this ordering or its reverse version contains at least half of the sum of all coefficients. So $\frac{1}{2}$-approximation of the LOP is trivial, but nothing better is known.

2.2 Construction Heuristics

We will now review some of the construction heuristics, i.e., methods which follow some principle for successively constructing a linear ordering. The principle should somehow reflect that we are searching for an ordering with high value.

2.2.1 Early LOP Methods

The Method of Chenery and Watanabe

One of the earliest heuristic methods was proposed by Chenery and Watanabe [37]. These authors did not formulate an algorithm, but just gave some ideas of how to obtain plausible rankings of the sectors of an input-output table. Their suggestion is to rank those sectors first which show a small share of inputs from other sectors and of outputs to final demand. Sectors having a large share of inputs from other industries and of final demand output should be ranked last. Chenery and Watanabe defined coefficients taking these ideas into account to find a preliminary ranking. Then they try to improve this ranking in some heuristic way which is not specified in their paper. The authors admit that their method does not necessarily lead to good approximate solutions of the triangulation problem.

Heuristics of Aujac & Masson

This method [7] is based on so-called *output coefficients*. The output coefficient of a sector i with respect to another sector j is defined as

$$b_{ij} = \frac{c_{ij}}{\sum\limits_{k \neq i} c_{ik}}.$$

Then it is intended to rank sector i before sector j whenever $b_{ij} > b_{ji}$ ("better customer principle"). This is impossible in general. So it is heuristically tried to find a linear ordering with few contradictions to this principle. Subsequently local changes are performed to achieve better triangulations. Similarly an *input coefficient* method can be formulated based on the input coefficients

$$a_{ij} = \frac{c_{ij}}{\sum\limits_{k \neq j} c_{kj}}.$$

Heuristics of Becker

In [12] two further methods are described. The first one is related to the previous ones in that it calculates special quotients to rank the sectors. For each sector i the number

$$q_i = \frac{\sum\limits_{k \neq i} c_{ik}}{\sum\limits_{k \neq i} c_{ki}}$$

is determined. The sector with the largest quotient q_i is then ranked highest. Its corresponding rows and columns are deleted from the matrix, and the procedure is applied to the remaining sectors.

Heuristic of Becker (1)

   (1) Set $S = \{1, 2, \ldots, n\}$.
   (2) For $k = 1, 2, \ldots, n$:

$$\text{(2.1) For each } i \in S \text{ compute } q_i = \frac{\sum\limits_{j \in S \setminus \{i\}} c_{ij}}{\sum\limits_{j \in S \setminus \{i\}} c_{ji}}.$$

     (2.2) Let $q_j = \max\{q_i \mid i \in S\}$.
     (2.3) Set $i_k = j$ and $S = S \setminus \{j\}$.

The second method starts with an arbitrarily chosen linear ordering, w.l.o.g. $\langle 1, 2, \ldots, n \rangle$. Then for every $m = 1, 2, \ldots, n - 1$ the objective function values of the orderings $\langle m+1, m+2, \ldots, n, 1, \ldots, m \rangle$ are evaluated. The best one among them is chosen, and the procedure is repeated as long as improvements are possible.

Heuristic of Becker (2)

   (1) Generate a random ordering.
   (2) Let $\langle i_1, i_2, \ldots, i_n \rangle$ denote the current ordering.
   (3) Evaluate all of the orderings $\langle i_{m+1}, i_{m+2}, \ldots, i_n, 1, 2, \ldots, i_m \rangle$, for $m = 1, 2, \ldots, n - 1$.
   (4) If the best one among these orderings is better than the current one, take it as the new current ordering and goto (3).

2.2.2 Reconstruction in the MDP

The first constructive method for the Maximum Diversity Problem was proposed by Ghosh in 1996 [66]. The author considered a multi-start algorithm consisting in a construction phase and a local search post-processing. In each iteration of the construction phase, one element is selected according to an estimation of its contribution to the final diversity, until the required m elements have been selected.

Glover et al. [68] also proposed simple constructive methods for the MDP in the late nineties. The authors considered that being the maximization of diversity a general model that can be customized to many different applications (for example by including additional constraints), it is important to propose a simple and general framework. To focus on a particular set of such constraints may not yield outcomes that can be readily extrapolated to constraints of a different structure. Therefore, they considered the MDP and give simple constructive methods that are easily adaptive.

An essential factor in their approach, is to characterize heuristics whose basic moves for transitioning from one solution to another are both simple and flexible allowing these moves to be adapted to multiple settings.

The consideration above led the authors in [68] to propose constructive and destructive methods for the MDP. Constructive methods are probably the first approach to try to solve a selection problem, in which a solution consists in selecting some elements out of a given set. In the particular case of the MDP, it is clear that a method that starting from an empty set M, iteratively selects elements, up to select m of them, is a straightforward approach. Glover et al. went one step further and coupled this naive procedure with a destructive method that removes elements from M. The alternation of both methods, constructing, destructing, and reconstructing iteratively, creates an oscillation pattern that gives the name to a method, Strategic Oscillation, that can be applied to other problems, which constitutes in this way a simple metaheuristic. The first constructive and destructive methods, C1 and D1 respectively, are based on the concept of center of gravity [69]. Given a set of points, X, the center of gravity in physics, also called the centroid or geometric center in mathematics, $center(X)$, is the arithmetic mean position of the points. It can be easily computed as the mean of the points coordinates. Algorithm C1 basically selects, at each step, the element i^* with the maximum distance to the center from among the elements already selected. The method finishes when m elements have been selected. On the contrary, starting with all the elements selected, D1 unselects, at each step, the element i^* with the minimum distance to the center from among the selected elements. The method finishes when $n - m$ elements have been unselected.

Let $G_n = (V_n, E_n)$ be the complete undirected graph on n nodes ($V_n = \{1, 2, \dots, n\}$), and let d_{ij} be the distance between nodes i and j (i.e., of edge $(i, j) \in E_n$). A feasible solution of the MDP is a set $M \subset V_n$ of m elements, and its value is the sum of distances between the pairs of nodes in M. As shown in the pseudo-code, Algorithm C1 first sets the partial solution under construction $M = \emptyset$, and the computes the centroid c of V_n. Then, it performs m iterations, adding in each one the element at a maximum distance to the centroid to M.

Constructive method C1

(1) Set $V_n = \{1, 2, \dots, n\}$.

(2) Set $M = \emptyset$.

(3) Compute $c = center(V_n)$

(4) For $k = 1, 2, \dots, m$:

    (4.1) Compute i^* such that $d_{i^* j} = max_{i \in V_n \setminus M} d_{ic}$.

    (4.2) $M = M \cup \{i^*\}$

    (4.3) $c = center(M)$

Destructive method D1 starts with the partial solution M with all the elements in V_n selected, thus being an unfeasible solution. Then, it performs $n - m$ iterations, removing in each one the element at a minimum distance to the centroid of M.

The second constructive and destructive heuristics, C2 and D2 respectively, are variations on the first ones presented above. Specifically, instead of constructing a center of the set, the methods consider the distance between an element i^* and the set, called the composite distance, simply as the sum of the distances between i^* and the elements in the set. In this way, C2 adds at each step the element i^* with maximum distance to the partial set M, and its counterpart, D2, removes at each step the element i^* with minimum distance to M.

Destructive method D1

(1) Set $M = V_n = \{1, 2, \ldots, n\}$.
(2) Compute $c = center(V_n)$
(3) For $k = m, m-1, \ldots, m-n$:

    (3.1) Compute i^* such that $d_{i^*j} = min_{i \in M} d_{ic}$.
    (3.2) $M = M \setminus \{i^*\}$
    (3.3) $c = center(M)$

C1, D1, C2, and D2 are of complexity orders $\mathcal{O}(mnr)$, $\mathcal{O}(n^2 r)$, $\mathcal{O}(nm^2)$, and $\mathcal{O}(n^3)$ respectively, where n is the number of given elements, m the number to be selected, and r the number of attributes defining each element.

2.2.3 General Insertions

We use the term insertion in different settings and its meaning can have subtle differences. In the context of a selection problem, such as the MDP, we may say that a constructive method inserts an element in the partial solution under construction. In that context, we could say that the constructive method adds an element to the solution as well, because we are not implying any order in the elements of the solution, and therefore we use insert or add indistinctly. However, in an ordering problem, such as the LOP, if we use the term insertion, it has a different connotation, because here it implies to insert in a specific position of an ordering. In this way, we have to specify in which position are we going to insert the element.

We describe now a simple heuristic for the LOP which builds an ordering by inserting the next objects at positions which are locally optimal.

Best Insertion for the LOP

(1) Select an arbitrary object j and set $S = \{1, 2, \ldots, n\} \setminus \{j\}$. Let $\langle j \rangle$ be the current ordering.

(2) For $k = 1, 2, \ldots, n - 1$:

    (2.1) Let $\langle i_1, i_2, \ldots, i_k \rangle$ denote the current ordering and choose some $l \in S$.

    (2.2) For every t, $1 \leq t \leq k + 1$, compute $q_t = \sum_{j=1}^{t-1} c_{i_j l} + \sum_{j=t}^{k} c_{l i_j}$ and let $q_p = \max\{q_t \mid 1 \leq t \leq k\}$.

    (2.3) Insert l at position p in the current ordering and set $S = S \setminus \{l\}$.

An alternative version of step (2.2) computes

$$q_t = \sum_{j=1}^{t-1} c_{i_j l} + \sum_{j=t}^{k} c_{l i_j} - \sum_{j=1}^{t-1} c_{l i_j} - \sum_{j=t}^{k} c_{i_j l}$$

to account for the sum of entries which are "lost" when l is inserted at position t.

2.2.4 Numerical Experiments

The Linear Ordering Problem

Table 2.1 reports on our results for 7 constructive heuristics on the OPT-I set (the set of 229 instances with optimum known). In this experiment we compute for each instance and each method the relative deviation *Dev* (in percent) between the best solution value *Value* obtained with the method and the optimal value for that instance. For each method, we also report the number of instances *#Opt* for which an optimum solution could be found. In addition, we calculate the so-called *score* statistic [147] associated with each method. For each instance, the *nrank* of method M is defined as the number of methods that produce a better solution than the one found by M. In the event of ties, the methods receive the same *nrank*, equal to the number of methods strictly better than all of them. The value of *Score* is the sum of the *nrank* values for all the instances in the experiment. Thus the lower the *Score* the better the method. We do not report running times in this table because these methods are very fast and their running times are extremely short (below 1 millisecond).

Specifically, Table 2.1 shows results for:

- CW: Chenery and Watanabe algorithm
- AM-O: Aujac and Masson algorithm (output coefficients)
- AM-I: Aujac and Masson algorithm (input coefficients)
- Bcq: Becker algorithm (based on quotients)
- Bcr: Becker algorithm (based on rotations)

Table 2.1 LOP constructive methods on OPT-I instances

	CW	AM–O	AM–I	Bcq	Bcr	BI1	BI2
IO							
Dev(%)	19.07	32.94	31.45	4.07	30.19	3.24	4.18
Score	231	291	266	101	289	89	104
#Opt	0	0	0	0	0	0	0
SGB							
Dev(%)	12.83	26.15	26.15	3.57	31.56	3.89	3.03
Score	100	125	125	54	175	56	40
#Opt	0	0	0	0	0	0	0
RandomAII							
Dev(%)	2.60	36.50	36.55	1.57	37.75	1.09	1.26
Score	100	135	136	68	162	34	48
#Opt	0	0	0	0	0	0	0
RandomB							
Dev(%)	10.13	24.69	24.69	7.04	26.41	5.24	4.87
Score	276	368	368	194	454	124	106
#Opt	0	0	0	0	0	0	0
MB							
Dev(%)	8.40	43.37	43.37	2.90	40.30	2.49	2.27
Score	120	178	178	80	154	52	48
#Opt	0	0	0	0	0	0	0
Special							
Dev(%)	0.02	0.57	0.14	3.10	0.40	0.01	0.17
Score	64	178	113	210	149	41	83
#Opt	0	0	0	0	0	4	3
OPT-I							
Avg. Dev(%)	10.85	32.97	32.55	3.95	32.35	3.49	3.50
Sum #Opt	0	0	0	0	0	4	3

- BI1: Best Insertion algorithm (variant 1)
- BI2: Best Insertion algorithm (variant 2)

Results in Table 2.1 clearly indicate that OPT-I instances pose a challenge for the simple heuristics with average percentage deviations ranging from 3.49% to 32.97%. In most of the cases none of the methods is able to match the optimum solution (with the exception of BI1 and BI2 with 4 and 3 optima respectively in the Special instances). These results show that only Bcq, BI1 and BI2 can be considered reasonable construction heuristics (with an average percent deviation lower than 5%).

The Maximum Diversity Problem

Table 2.2 reports the experiments in [69] on the GKD-a instances. The authors computed the optimal solutions of the instances in this set with a specific purpose method since they found that the mathematical programming model was unable to solve them. The table reports the average percentage deviation of the heuristic solutions obtained with constructive methods C1, D1, C2, and D2 respectively on each subset according to the size.

The overall average percentage deviations of the constructive methods are 1.92 (C1), 0.52 (D1), 0.54 (C2), and 0.53 (D2). On the other hand, the number of instances in which each method is able to match the optimal value out of the 75 instances in this set is: 46 (C1), 67 (D1), 69 (C2), and 73 (D2). This seems to indicate that D2 and C2 are the best methods, which is an interesting result considering that D2 is a destructive method, based on removing the bad elements from the solution, and C2 is a constructive one, based on adding the good elements to the solution. Additionally, this table also shows that the performance of the method heavily relies on the value of m. For example, C1 exhibits an average percentage deviation to the optimal value of 8.20 on instances with $n = 15$ and $m = 2$, but this deviation drops to 0.00 on instances with $n = 15$ and $m = 6$.

It must be noted that the instances in the GKD-a set are very small. They were actually considered of that size to easily compute their optimal solutions. However, as will be shown in the next chapter, these methods quickly deteriorate when the size increases. This is why we would need more elaborated metaheuristics to target real size instances.

Table 2.2 MDP constructive methods

	C1	D1	C2	D2
n = 10				
$m = 2$	6.68	1.63	2.16	1.63
$m = 3$	3.90	1.01	0.86	1.01
$m = 4$	2.97	1.41	1.18	0.04
$m = 6$	0.78	0.00	0.00	0.00
$m = 8$	0.10	0.10	0.07	0.00
n = 15				
$m = 2$	8.20	0.88	1.20	4.10
$m = 4$	0.72	0.00	0.79	0.00
$m = 6$	0.00	0.00	0.00	0.00
$m = 9$	0.39	0.70	0.09	0.01
$m = 11$	0.34	0.34	0.00	0.01
n = 30				
$m = 5$	1.79	0.12	1.02	0.59
$m = 8$	0.46	0.46	0.02	0.02
$m = 12$	0.33	0.48	0.11	0.05
$m = 18$	0.15	0.10	0.02	0.00

2.3 Local Search

After having constructed a solution with one of the heuristics above, it is reasonable
to look for improvement possibilities. In this section we will describe fairly simple
(deterministic) local improvement methods that are able to produce acceptable solutions for the LOP and the MDP. The basic philosophy that drives local search is that
it is often possible to find a good solution by repeatedly increasing the quality of a
given solution, making small changes at a time called *moves*. The different types of
possible moves characterize the various heuristics. Starting from a solution generated by a construction heuristic, a typical local search performs steps as long as the
objective function increases (since we are considering maximization problems). Let
first provide some basic notions in optimization to better understand the importance
and implications of a move definition.

In general terms, an optimization problem consists of finding the best (optimal)
feasible solution in a space defined by the solution representation and the problem
constraints. In the LOP we search in a space defined by all permutations of n elements, while in the MDP we search for a selection of m elements out of n elements.
In the permutation representation, a standard move is an insertion of an element
from one position to another one. In the binary representation used for the MDP,
a typical move entails *switching* one of the binary variables from its current value
to its complementary value, that is, from zero to one or from one to zero, which is
equivalent to un-select one selected element, and select an un-selected one.

A *local optimal* solution in the search space is one for which there is no neighboring solution with a better objective function value, where the neighboring solutions
are all those that can be reached in a single move. A *global optimal* solution, on the
other hand, is such that no other solution in the search space is better than it.

An important concept when defining a move, and thus its associated neighborhood, is *connectivity*, which refers to whether or not there exists at least one path of
moves that will connect all pairs of solutions in the solution space. For instance, if
the representation is a binary vector and the move mechanism to explore the solution
space consists of switching the value of a single variable then it can be shown that
all pairs of solutions are connected. In other words, any solution can be transformed
into any other one by changing one value at a time.

Local search can only be expected to obtain optimal or near-optimal solutions for
easy problems of medium size, but it is a very important and powerful concept for
the design of meta-heuristics, which are the topic of the next chapter.

2.3.1 Insertions in permutation problems

This heuristic checks whether the objective function can be improved if the position
of an object in the current ordering is changed. All possibilities for altering the position of an object are checked and the method stops when no further improvement
is possible this way.

In problems where solutions are represented as permutations, insertions are probably the most direct and efficient way to modify a solution. Note that other movements, such as swaps, can be obtained by composition of two or more insertions. We define $move(O_j, i)$ as the modification which deletes O_j from its current position j in permutation O and inserts it at position i (i.e., between the objects currently in positions $i - 1$ and i).

Now, the insertion heuristic tries to find improving moves examining eventually all possible new positions for all objects O_j in the current permutation O. There are several ways for organizing the search for improving moves. For our experiments we proceeded as follows:

Insertion

(1) Compute an initial permutation $O = \langle O_1, O_2, \ldots, O_n \rangle$.

(2) For $j = 1, 2, \ldots, n$:

    (2.1) Evaluate all possible insertions $move(O_j, i)$.

    (2.2) Let $move(O_k, i^*)$ be the best of these moves.

    (2.3) If $move(O_k, i^*)$ is improving then perform it and update O.

(3) If some improving move was found, then goto (2).

In [106] two neighborhoods are studied in the context of local search methods for the LOP. The first one consists of permutations obtained by switching the positions of contiguous objects O_j and O_{j+1}. The second one involves all permutations resulting from executing general insertion moves, as defined above. The conclusion from the experiments is that the second neighborhood clearly outperforms the first one, which is much more limited. Furthermore two strategies for exploring the neighborhood of a solution were studied. The *best* strategy selects the move with the largest *move value* among all the moves in the neighborhood. The *first* strategy, on the other hand, scans the list of objects (in the order given by the current permutation) searching for the first object whose movement gives a strictly positive move value. The computations revealed that both strategies provide similar results but the *first* involved lower running times.

k-opt

A generalization of the insertion heuristic is to simultaneously change the position of several elements in the ordering. We may remove several elements from the solution, inserting them in a different position, and then evaluate if the whole operation improves the solution. The most popular method implementing such a move is the k-opt, originally applied to the traveling salesman problem, following a principle that can be applied to many combinatorial optimization problems. Basically, it selects k elements of a solution and locally optimizes with respect to these elements. For the LOP, a possible k-opt heuristic would be to consider all subsets of k objects

$O_{i_1}, \ldots, O_{i_k}$ in the current permutation and find the best insertion of these objects to the positions $i_1, \ldots, i_k$. Since the number of possible new assignments grows exponentially with k, it is usually implemented for $k = 2$ (2-opt) and $k = 3$ (3-opt).

2.3.2 Exchanges in binary problems

Solutions to selection problems, such as the MDP, are typically represented with binary variables. Note that we could also consider an alternative representation consisting of the identities of the selected elements. A solution would then be represented by m discrete values in the range from 1 to n. We usually called it the *integer representation*, as opposite to the *binary representation*.

The first local search method proposed for the MDP was due to Ghosh [66] in 1996. It implements a straightforward *hill climbing* heuristic based on performing the best available exchange. Exchanges in this context consist of replacing one selected element with an unselected one. In mathematical terms, we denote a solution with the set M of selected elements, and its objective function value (sum of pairwise distances) as $z(M)$. If we replace $i \in M$ with $j \in V_n \setminus M$, we obtain a new solution M', and its objective value can be incrementally computed from $z(M)$ with the expression $z(M') = z(M) + value(i, j)$, where the so-called move-value, $value(i, j)$, is computed as:

$$value(i, j) = \sum_{k \in M \setminus \{i\}} d_{jk} - d_{ik}.$$

The procedure scans the set of selected elements M in search for the exchange with the largest move-value (i.e., with the largest increase of the objective function). The method performs the best available exchange in each iteration until no further improvement is possible.

This early local search implementation for the MDP, LS, performs an exhaustive exploration of the neighborhood, in which it scans the elements following an arbitrary ordering. As pointed out by other researchers on diversity problems (see for instance Silva et al. [153]), this method produces results of relatively good quality but it presents very long running times. It is nowadays well-documented that an efficient implementation of a local search method should include strategies to speed up the identification of good moves. In general terms, we refer them as *candidate list strategies*.

Duarte and Martí [51] proposed a first improvement to increase the efficiency of LS. Specifically, the authors define the contribution D_i of element $i \in M$ to the objective value $z(M)$ as:

$$D_i = \sum_{k \in M} d_{ik}.$$

Instead of computing the move value $value(i, j)$ for all $i \in M$ and $j \in V_n \setminus M$, this improved local search, ILS, first selects the element i^* in M with the lowest contribution to the value of the current solution. Then, instead of scanning the whole set $V_n \setminus M$ searching for the best exchange associated with i^*, ILS restricts itself to performing the first improving move (without examining the remaining elements in $V_n \setminus M$). If there is no improving move associated with i^*, the method resorts to the next element with the lowest D_i value and so on. This improved local search method performs iterations until no further improvement is possible.

As will be shown in Chapter 3, different strategies and enhancements have been proposed to create an efficient local search for the MDP. As a matter of fact, the resulting advanced local searches constitute the core of many metaheuristics for the MDP. This is especially evident in the paper entitled *"Comparing local search metaheuristics for the maximum diversity problem"* [8], which will be reviewed in the next chapter.

2.3.3 LOP methods

We now describe some heuristics specifically designed to the LOP. They are based on properties of this problem.

The Heuristic of Chanas & Kobylanski

The method developed by Chanas and Kobylanski [37], referred to as the CK method in the following, is based on the following symmetry property of the LOP. If the permutation $O = \langle O_1, O_2, \ldots, O_n \rangle$ is an optimum solution to the maximization problem, then an optimum solution to the minimization problem is $O^* = \langle O_n, O_{n-1}, \ldots, O_1 \rangle$. In other words, when the sum of the elements above the main diagonal is maximized, the sum of the elements below the diagonal is minimized. The CK method utilizes this property to escape local optimality. In particular, once a local optimum solution O is found, the process is re-started from the permutation O^*. This is called the REVERSE operation.

In a global iteration, the CK method performs insertions as long as the solution improves. Given a solution, the algorithm explores the insertion move $move(O_j, i)$ of each element O_j in all the positions i in O, and performs the best one. When no further improvement is possible, it generates a new solution by applying the REVERSE operation from the last solution obtained, and performs a new global iteration. The method finishes when the best solution found cannot be improved upon in the current global iteration.

It should be noted that the CK method can be considered to be a generalization of the second heuristic of Becker described above. The latter evaluates the orderings that can be obtained by rotations of a solution, while the CK method evaluates all insertions. Since these rotations are basically insertions of the first elements to the last positions, we can conclude that Becker's method explores only a fraction of the solutions explored by CK.

Kernighan-Lin Type Improvement

The main problem with local improvement heuristics is that they very quickly get trapped in a local optimum. Kernighan and Lin proposed the idea (originally in [97] for a partitioning problem) of looking for more complicated moves that are composed of simpler moves. In contrast to pure improvement heuristics, it allows that some of the simple moves are not improving. In this way the objective can decrease locally, but new possibilities arise for escaping from the local optimum. This type of heuristic proved particularly effective for the traveling salesman problem (where it is usually named *Lin-Kernighan heuristic*).

We only describe the principle of the Kernighan-Lin approach. For practical applications on large problems, it has to be implemented carefully with appropriate data structures and further enhancements like restricted search or limited length of combined moves to speed up the search for improving moves. We do not elaborate on this here.

We devised two heuristics of this type for the LOP. In the first version, the basic move consists of interchanging two objects in the current permutation.

Kernighan-Lin 1

(1) Compute some linear ordering O.

(2) Let $m = 1$, $S_m = \{1, 2, \ldots, n\}$.

(3) Determine objects $s, t \in S_m$, $s \neq t$, the interchange of which in the current ordering leads to the largest increase g_m of the objective function (increase may be negative).

(4) Interchange s and t in the current ordering. Set $s_m = s$ and $t_m = t$.

(5) If $m < \lfloor n/2 \rfloor$, set $S_{m+1} = S_m \setminus \{s, t\}$ and $m = m + 1$. Goto (3).

(6) Determine $1 \leq k \leq m$, such $G = \sum_{i=1}^{k} g_i$ is maximum.

(7) If $G \leq 0$ then Stop, otherwise, starting from the original ordering O, successively interchange s_i and t_i, for $i = 1, 2, \ldots, k$. Let O denote the new ordering and goto (2).

The second version builds upon insertion moves.

Kernighan-Lin 2

(1) Compute some linear ordering O.

(2) Let $m = 1$, $S_m = \{1, 2, \ldots, n\}$.

(3) Among all possibilities for inserting an object of S_m at a new position determine the one leading to the largest increase g_p of the objective function (increase may be negative). Let s be this object and p the new position.

(4) Move s to position p in the current ordering. Set $s_m = s$ and $p_m = p$.

(5) If $m < n$, set $S_{m+1} = S_m \setminus \{s\}$ and $m = m + 1$. Goto (3).

(6) Determine $1 \leq k \leq m$, such $G = \sum_{i=1}^{k} g_i$ is maximum.

(7) If $G \leq 0$ then Stop, otherwise, starting from the original ordering O, successively move s_i to position p_i, for $i = 1, 2, \ldots, k$. Let O denote the new ordering and goto (2).

Local Enumeration

This heuristic chooses windows $\langle i_k, i_{k+1}, \ldots, i_{k+L-1} \rangle$ of a given length L of the current ordering $\langle i_1, i_2, \ldots, i_n \rangle$ and determines the optimum subsequence of the respective objects by enumerating all possible orderings. The window is moved along the complete sequence until no more improvements can be found. Of course, L cannot be chosen too large because the enumeration needs time $O(L!)$.

Local Enumeration

(1) Compute some linear ordering O.

(2) For $i = 1, \ldots, n - L + 1$:

   (2.1) Find the best possible rearrangement of the objects at positions $i, i+1, \ldots, i+L-1$.

(3) If an improving move has been found in the previous loop, then goto (2).

Table 2.3 reports on our results for 7 improving heuristics on the OPT-I set of instances. As in the construction heuristics, we report, for each instance and each method, the relative percent deviation *Dev*, the number of instances *#Opt* for which an optimum solution is found, and the *score* statistic. Similarly, we do not report running times in this table because these methods are fairly fast. Specifically, the results obtained with the following improvement methods (started with a random initial solution) are given:

–   LSi: Local Search based on insertions

–   2opt: Local Search based on 2-opt

- 3opt: Local Search based on 3-opt
- LSe: Local Search based on exchanges
- KL1: Kernighan-Lin based on exchanges
- KL2: Kernighan-Lin based on insertions
- LE: Local enumeration

Table 2.3 Improvement methods on OPT-I instances

	LSi	2opt	3opt	LSe	KL1	KL2	LE
IO							
Dev(%)	1.08	0.64	0.23	1.73	1.35	4.24	0.01
Score	243	181	125	295	239	232	49
#Opt	0	1	4	0	1	0	43
SGB							
Dev(%)	0.16	0.81	0.53	1.35	0.63	0.28	1.09
Score	42	122	84	154	100	63	135
#Opt	1	0	0	0	0	0	0
RandomAII							
Dev(%)	0.16	0.77	0.38	0.62	0.61	0.09	0.54
Score	46	161	81	134	134	29	112
#Opt	0	0	0	0	0	0	0
RandomB							
Dev(%)	0.79	4.04	2.13	3.78	3.51	0.61	3.56
Score	124	400	232	387	359	95	362
#Opt	1	0	0	0	0	1	0
MB							
Dev(%)	0.02	0.57	0.14	3.10	0.40	0.01	0.17
Score	64	178	113	210	149	41	83
#Opt	0	0	0	0	0	4	3
Special							
Dev(%)	1.19	3.30	2.05	3.21	2.40	0.89	3.52
Score	69	144	82	138	120	49	156
#Opt	4	2	2	2	3	3	3
OPT-I							
Avg. Dev(%)	0.57	1.69	0.91	2.30	1.49	1.02	1.48
Sum #Opt	5	3	6	2	4	8	49

As expected, the improvement methods are able to obtain better solutions than the construction heuristics, with average percentage deviations (shown in Table 2.3) ranging from 0.57% to 2.30% (the average percentage deviations of the construction heuristics range from 3.49% to 32.97% as reported in Table 2.1). We have not observed significant differences when applying the improvement method from different initial solutions. For example, as shown in Table 2.3 the LSi method exhibits a *Dev* value of 0.16% on the RandomAII instances when it is started from random solutions. When it is run from the CW or the Bcr solutions, it obtains a *Dev* value of 0.17% and 0.18% respectively.

2.4 Multi-Start Procedures

Multi-start procedures were originally conceived as a way of exploiting a local or neighborhood search procedure, by simply applying it from multiple random initial solutions. It is well known that search methods based on local optimization, aspiring to find global optima, usually require certain diversification to overcome local optimality. Without this diversification, such methods can become reduced to tracing paths that are confined to a small area of the solution space, making it impossible to find a global optimum. *Multi-start algorithms* can be considered to be a bridge between simple (classical) heuristics and complex (modern) meta-heuristics. The *re-start mechanism* of multi-start methods can be super-imposed on many different search methods. Once a new solution has been generated, a variety of options can be used to improve it, ranging from a simple greedy routine to a complex meta-heuristic. This section focuses on the different strategies and methods that can be used to generate solutions to launch a succession of new searches for a global optimum.

The principle layout of a multi-start procedure is the following.

Multi-Start

(1) Set $i=1$.
(2) While the stopping condition is not satisfied:

    (2.1) Construct a solution x_i. *(Generation)*
    (2.2) Apply local search to improve x_i and let x_i' be the solution obtained. *(Improvement)*
    (2.3) If x_i' improves the best solution, update it. Set $i = i+1$. *(Test)*

The computation of x_i in (2.1) is typically performed with a constructive algorithm. Step (2.2) tries to improve this solution, obtaining x_i'. Here, a simple improvement method can be applied. However, this second phase has recently become more elaborate and, in some cases, is performed with a complex meta-heuristic that may or may not improve the initial solution x_i (in this latter case we set $x_i' = x_i$).

2.4.1 Variants of Multi-Start

We will first review some relevant contributions on multi-start procedures.

Early papers on multi-start methods are devoted to the Monte Carlo random re-start in the context of nonlinear unconstrained optimization, where the method simply evaluates the objective function at randomly generated points. The probability of success approaches 1 as the sample size tends to infinity under very mild assumptions about the objective function. Many algorithms have been proposed that combine the Monte Carlo method with local search procedures [148]. The convergence for random re-start methods is studied in [158], where the probability distribution used to choose the next starting point can depend on how the search evolves. Some extensions of these methods seek to reduce the number of complete local searches that are performed and increase the probability that they start from points close to the global optimum [123].

In [17] relationships among local minima from the perspective of the best local minimum are analyzed, finding convex structures in the cost surfaces. Based on the results of that study, they propose a multi-start method where starting points for greedy descent are adaptively derived from the best previously found local minima. In the first step, *adaptive multi-start heuristics* generate random starting solutions and run a greedy descent method from each one to determine a set of corresponding random local minima. In the second step, *adaptive starting solutions* are constructed based on the local minima obtained so far and improved with a greedy descent method. This improvement is applied several times from each adaptive starting solution to yield corresponding *adaptive local minima*. The authors test this method for the traveling salesman problem and obtain significant speedups over previous multi-start implementations.

Simple forms of multi-start methods are often used to compare other methods and measure their relative contribution. In [10] different genetic algorithms for six sets of benchmark problems commonly found in the genetic algorithms literature are compared: traveling salesman problem, job-shop scheduling, knapsack and bin packing problem, neural network weight optimization, and numerical function optimization. The author uses the *multi-start method*, also called *multiple restart stochastic hill-climbing* as a basis for computational testing. Since solutions are represented with strings, the improvement step consists of a local search based on random flipping of bits. The results indicate that using genetic algorithms for the optimization of static functions does not yield a benefit, in terms of the final answer obtained, over simpler optimization heuristics.

One of the most well known multi-start methods is the *greedy adaptive search procedure (GRASP)*. The GRASP methodology was introduced by Feo and Resende [57] and was first used to solve set covering problems [56]. We will devote a section in the next chapter to describe this methodology in detail.

A multi-start algorithm for unconstrained global optimization based on *quasi-random samples* is presented in [86]. Quasi-random samples are sets of deterministic points, as opposed to random, that are evenly distributed over a set. The algorithm applies an inexpensive local search (steepest descent) on a set of quasi-random points to concentrate the sample. The sample is reduced, replacing worse points with new quasi-random points. Any point that is retained for a certain number of iterations is used to start an efficient complete local search. The algorithm terminates when no new local minimum is found after several iterations. An experimental comparison shows that the method performs favorably with respect to other global optimization procedures.

An open question in order to design a good search procedure is whether it is better to implement a simple improving method that allows a great number of global iterations or, alternatively, to apply a complex routine that significantly improves a few generated solutions. A simple procedure depends heavily on the initial solution but a more elaborate method takes much more running time and therefore can only be applied a few times, thus reducing the sampling of the solution space. Some meta-heuristics, such as GRASP, launch limited local searches from numerous constructions (i.e., starting points). In other methods, such as tabu search, the search starts from one initial point and, if a restarting procedure is also part of the method, it is invoked only a limited number of times. In [118] the balance between restarting and search-depth (i.e., the time spent searching from a single starting point) is studied in the context of the matrix bandwidth problem. Both alternatives were tested with the conclusion that it was better to invest the time searching from a few starting points than re-starting the search more often. Although we cannot draw a general conclusion from these experiments, the experience in the current context and in previous projects indicates that some meta-heuristics, like tabu search, need to reach a critical search depth to be effective. If this search depth is not reached, the effectiveness of the method is severely compromised.

2.4.2 Numerical Experiments

As done in previous sections, we will now apply the most relevant methods to solve our problems. The first subsection describes 10 multi-start methods proposed for the LOP, and the second subsection shows how a constructive method improves when it is adapted to a multi-start scheme in the case of the MDP.

The Linear Ordering Problem

In this section we will describe and compare 10 different constructive methods for the LOP. It should be noted that, if a constructive method is completely deterministic (with no random elements), its replication (running it several times) will always produce the same solution. Therefore, we should add random selections in a constructive method to obtain different solutions when replicated. Alternatively, we can modify selections from one construction to another in a deterministic way by recording and using some frequency information. We will look at both approaches, which will enable us to design constructive methods for the LOP that can be embedded in a multi-start procedure.

Above we have described the construction heuristic of Becker [12] in which for each object i the value q_i is computed. Then, the objects are ranked according to the q-values $q_i = \sum_{k \neq i} c_{ik} / \sum_{k \neq i} c_{ki}$.

We now compute two other values that can also be used to measure the attractiveness of an object to be ranked first. Specifically, r_i and c_i are, respectively, the sum of the elements in the row corresponding to object i, and the sum of the elements in the column of object i, i.e., $r_i = \sum_{k \neq i} c_{ik}$ and $c_i = \sum_{k \neq i} c_{ki}$.

Constructive Method G1

This method first computes the r_i values for all objects. Then, instead of selecting the object with the largest r-value, it creates a list with the most attractive objects, according to the r-values, and randomly selects one among them. The selected object is placed first and the process is repeated for n iterations. At each iteration the r-values are updated to reflect previous selections (i.e., we sum the c_{ik} across the unselected elements) and the candidate list for selection is computed with the highest evaluated objects. The method combines the *random selection* with the *greedy evaluation*, and the size of the candidate list determines the relative contribution of these two elements.

Constructive method G1

(1) Set $S = \{1, 2, \ldots, n\}$. Let $\alpha \in [0, 1]$ be the percentage for selection and O be the empty ordering.

(2) For $t = 1, 2, \ldots, n$:

   (2.1) Compute $r_i = \sum_{k \in S, k \neq i} c_{ik}$ for all $i \in S$.

   (2.2) Let $r^* = \max\{r_i \mid i \in S\}$.

   (2.3) Compute the candidate list $C = \{i \in S \mid r_i \geq \alpha r^*\}$.

   (2.4) Randomly select $j^* \in C$ and place j^* at position t in O and set $S = S \setminus \{j^*\}$.

Constructive Methods G2 and G3

Method G2 is based on the c_i-values computed above. It works in the same way as G1 but the attractiveness of object i is now measured with c_i instead of r_i. Objects with large c-values are placed now in the last positions.

Constructive method G2

(1) Set $S = \{1, 2, \ldots, n\}$. Let $\alpha \in [0, 1]$ be the percentage for selection and O be the empty ordering.

(2) For $t = n, n-1, \ldots, 1$:

    (2.1) Compute $c_i = \sum\limits_{k \in S, k \neq i} c_{ki}$ for all $i \in S$.

    (2.2) Let $c^* = \max\{c_i \mid i \in S\}$.

    (2.3) Compute the candidate list $C = \{i \in S \mid c_i \geq \alpha c^*\}$.

    (2.4) Randomly select $j^* \in C$ and place j^* in position t in O and set $S = S \setminus \{j^*\}$.

In a similar way, constructive method G3 measures the attractiveness of object i for selection with q_i and performs the same steps as G1. Specifically, at each iteration the q-values are computed with respect to the unselected objects, a restricted candidate list is formed with the objects with largest q-values, and one of them is randomly selected and placed first.

Constructive Methods G4, G5 and G6

These methods are designed analogously to G1–G3, except that the selection of objects is from a candidate list of the least attractive and the solution is constructed starting from the last position of the permutation. We give the specification of G6 which is modification of G3.

Constructive method G6

(1) Set $S = \{1, 2, \ldots, n\}$. Let $\alpha \geq 0$ be the percentage for selection and O be the empty ordering.

(2) For $t = 1, 2, \ldots, n$:

    (2.1) For all $i \in S$, compute

$$q_i = \frac{\sum\limits_{k \in S, k \neq i} c_{ik}}{\sum\limits_{k \in S, k \neq i} c_{ki}}.$$

    (2.2) Let $q^* = \min\{q_i \mid i \in S\}$.

    (2.3) Compute the candidate list $C = \{i \in S \mid q_i \leq (1 + \alpha)q^*\}$.

    (2.4) Randomly select $j^* \in C$ and place j^* in position $n - t + 1$ in O and set $S = S \setminus \{j^*\}$.

Constructive Method MIX

This is a mixed procedure derived from the previous six. The procedure generates a fraction of solutions from each of the previous six methods and combines these solutions into a single set. That is, if n solutions are required, then each method Gi, $i = 1, \ldots, 6$, contributes $n/6$ solutions.

Constructive Method RND

This is a random generator. This method simply generates random permutations. We use it as a basis for our comparisons.

Constructive Method DG

This is a general purpose diversification generator suggested in [72] which generates diversified permutations in a systematic way without reference to the objective function.

Constructive Method FQ

This method implements an algorithm with frequency-based memory, as proposed in tabu search [69] (we will see this methodology in the next chapter). It is based on modifying a measure of attractiveness with a frequency measure that discourages objects from occupying positions that they have frequently occupied in previous solution generations.

The constructive method FQ (proposed in [24]) is based on the notion of constructing solutions employing modified *frequenciesfrequency*. The generator exploits the permutation structure of a linear ordering. A frequency counter is maintained to record the number of times an element i appears in position j. The frequency counters are used to penalize the "attractiveness" of an element with respect to a given position. To illustrate this, suppose that the generator has created 30 solutions. If 20 out of the 30 solutions have element 3 in position 5, then the frequency counter $freq(3,5) = 20$. This frequency value is used to bias the potential assignment of element 3 in position 5 during subsequent constructions, thus inducing *diversification* with respect to the solutions already generated.

The attractiveness of assigning object i to position j is given by the greedy function $fq(i,j)$, which modifies the value of q_i to reflect previous assignments of object i to position j, as follows:

$$fq(i,j) = \frac{\sum\limits_{k \neq i} c_{ik}}{\sum\limits_{k \neq i} c_{ki}} - \beta \frac{\max_q}{\max_f} freq(i,j),$$

where $\max_f = \max\{freq(i,j) \mid i = 1,\dots,n, j = 1,\dots,n\}$ and $\max_q = \max\{q_i \mid i = 1,\dots,n\}$.

Constructive method FQ

(1) Set $S = \{1,2,\dots,n\}$. Let $\beta \in [0,1]$ be the percentage for diversification and $freq(i,j)$ be the number of times object i has been assigned to position j in previous constructions.

(2) For $t = 1,2,\dots,n$:

(2.1) For all $i,j \in S$ compute $fq(i,j) = \dfrac{\sum\limits_{k \neq i} c_{ik}}{\sum\limits_{k \neq i} c_{ki}} - \beta \dfrac{\max_q}{\max_f} freq(i,j)$.

(2.2) Let i^* and j^* be such that $fq(i^*,j^*) = \max\{fq(i,j) \mid i,j \in S\}$.

(2.3) Place i^* at position j^* in O and set $S = S \setminus \{i^*\}$.

(2.4) $freq(i^*,j^*) = freq(i^*,j^*) + 1$.

It is important to point out that $fq(i,j)$ is an adaptive function since its value depends on attributes of the unassigned elements at each iteration of the construction procedure.

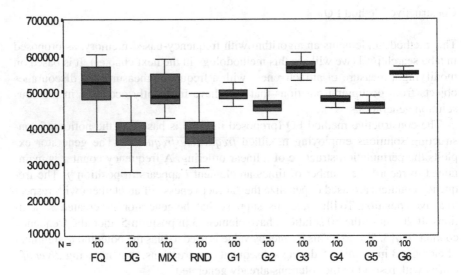

Fig. 2.1 Objective function value box-plot for each method

In our first experiment we use the instance `stabu75` from LOLIB. We have generated a set of 100 solutions with each of the 10 generation methods. Figure 2.1 shows in a box-and-whisker-plot representation, the value of the 100 solutions generated with each method. Since the LOP is a maximization problem, it is clear that the higher the value, the better the method. We can therefore say that constructive method G3 obtains the best results. Other methods, such as FQ and MIX also obtain solutions with very good values, but their box-plot representation indicates that they also produce lower quality solutions. However, if the construction is part of a global method (as is the case in multi-start methods), we may prefer a constructive method able to obtain solutions with different structures rather than a constructive method that provides very similar solutions. Note that if every solution is subjected to local search, then it is preferable to generate solutions scattered in the search space as starting points for the local search phase rather than solutions concentrated in the same area of the solution space. Therefore, we need to establish a trade off between *quality* and *diversity* when selecting our construction method.

Given a set of solutions P represented as permutations, in [120] a diversity measure d is proposed which consists of computing the distances between each solution and a "center" of P. The sum (or alternatively the average) of these $|P|$ distances provides a measure of the diversity of P. The diversity measure d is calculated as follows:

(1) Calculate the median position of each element i in the solutions in P.
(2) Calculate the *dissimilarity* (distance) of each solution in the population with respect to the median solution. The dissimilarity is calculated as the sum of the absolute difference between the position of the elements in the solution under consideration and the median solution.
(3) Calculate d as the sum of all the individual dissimilarities.

For example, assume that P consists of the orderings $\langle A,B,C,D \rangle$, $\langle B,D,C,A \rangle$, and $\langle C,B,A,D \rangle$. The median position of element A is therefore 3, since it occupies positions 1, 3 and 4 in the given orderings. In the same way, the median positions of B,C and D are 2, 3 and 4, respectively. Note that the median positions might not induce an ordering, as in the case of this example. The diversity value of the first solution is then calculated as $d_1 = |1-3| + |2-2| + |3-3| + |4-4| = 2$.

In the same way, the diversity values of the other two solutions are obtained as $d_2 = 4$ and $d_3 = 2$. The diversity measure d of P is then given by $d = 2+4+2 = 8$.

We then continue with our experiment to compare the different constructive methods for the LOP. As described above, we have generated a set of 100 solutions with each of the 10 generation methods. Figure 2.2 shows the box-and-whisker plot of the diversity values of the solution set obtained with each method.

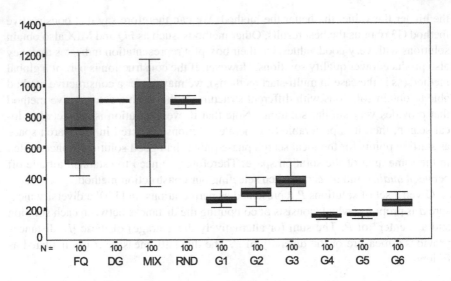

Fig. 2.2 Diversity value box-plot for each method

Figure 2.2 shows that MIX and FQ obtain the highest diversity values (but also generate other solutions with low diversity values). As expected, the random constructive method RND consistently produces high diversity values (always generating solutions with an associated d-value over 800 in the diagram).

As mentioned, a good method must produce a set of solutions with high quality and high diversity. If we compare, for example, generators MIX and G3 we observe in Fig. 2.1 that G3 produces slightly better solutions in terms of solution quality, but Fig. 2.2 shows that MIX outperforms G3 in terms of diversity. Therefore, we will probably consider MIX as a better method than G3. In order to rank the methods we have computed the average of both measures across each set.

Figure 2.3 shows the average of the diversity values on the x-axis and the average of the quality on the y-axis. A point is plotted for each method. As expected, the random generator RND produces a high diversity value (as measured by the dissimilarity) but a low quality value. DG matches the diversity of RND using a systematic approach instead of randomness, but as it does not use the value of solutions, it also presents a low quality score. The mixed method MIX provides a good balance between dissimilarity and quality, by uniting solutions generated with methods G1 to G6.

We think that quality and diversity are equally important, so we have added both averages. To do so, we use two relative measures $\triangle C$ for quality, and $\triangle d$ for diversity. They are basically standardizations to translate the average of the objective function values and diversity values respectively to the [0,1] interval. In this way we can simply add both quantities.

Figure 2.4 clearly shows the following ranking of the 10 methods, where the overall best is the FQ generator: G5, G4, G2, G1, DG, RND, G6, G3, MIX and FQ.

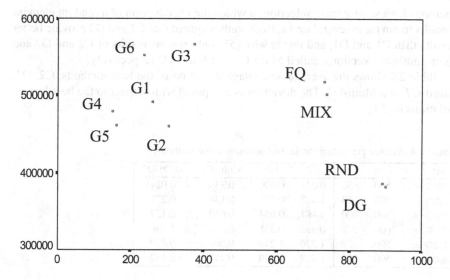

Fig. 2.3 Quality and diversity for each method

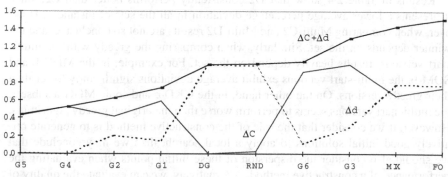

Fig. 2.4 Quality plus diversity for each method

These results are in line with previous works which show the inclusion of memory structures (frequency information) to be effective within the multi-start framework. However, one should note that this method of ranking has been obtained considering both quality and diversity with equal weight. If we vary this criterion, the ranking would also change.

The Maximum Diversity Problem

Duarte and Martí [51] implemented in C language the constructive methods C1, C2, D1, and D2 proposed in [69], and compared the original algorithms with multi-start versions of them. In particular, the authors replaced the greedy selection in these methods for a semi-greedy selection in which the introduction of a random element permits to run them several times. Both studies agreed that C2 and D2 provide better results than C1 and D1, and this is why [51] only reports results of C2 and D2 and their multi-start versions, called Multi-C2 and Multi-D2 respectively.

Table 2.4 shows the average percentage deviation of the four methods, C2, D2, Multi-C2, and Multi-D2. The deviation is computed with respect to the best-known solutions in [51].

Table 2.4 Average percentage deviations of constructive methods.

Set	n	m	C2	D2	Multi-C2	Multi-D2
GKD-c	500	50	0.021	0.006	0.336	0.024
MDG-a	500	50	1.027	0.729	0.174	0.277
MDG-a	500	200	0.487	0.034	0.630	0.127
MDG-a	2000	200	0.392	0.258	0.447	1.586
MDG-b	500	50	1.270	1.219	0.207	0.470
SOM-b	500	200	1.722	1.079	0.297	0.162

Results in Table 2.4 show that D2 consistently performs better than C2, since it presents a lower average percentage deviation in all the sets of instances. However, when comparing Multi-C2 and Multi-D2 results are not so conclusive, and the winner depends on the set. Similarly, when comparing the greedy with the multi-start versions, results heavily depend on the set. For example, in the MDG-b and SOM-b, the multi-start versions exhibit average deviations significantly lower than their greedy versions. On the other hand, in the GKD-c and some MDG-a subsets, the multi-start versions seems to perform worse than the original greedy algorithms. However, if we consider that the role of the constructive method is to generate relatively good initial solutions to apply a local search, then we may conclude that we should also consider the dispersion of these initial points when evaluating the performance of a constructive method. Alternatively, we can evaluate the quality of the solutions obtained after the application of the local search, to have a complete picture of the performance of the method.

Duarte and Martí [51] performed a second experiment comparing the multi-start methods coupled with the two local searches, LS and ILS, described above. In particular, the author consider two versions of the multi-start method applied to C2, called Multi-C2-LS and Multi-C2-ILS. Table 2.5 reports the average percent deviation of the Multi-C2 method, where no local search is applied, and these two methods in which a local search improves each constructed solution. The three methods are run for 10 seconds on each instance.

When comparing the results of the three methods reported in Table 2.5, we have to keep in mind that during the execution of each method for 10 seconds, Multi-C2 is able to generate a large number of solutions (roughly speaking around 1,000 on average), while Multi-C2-LS generates and improves a fraction of them (approximately 100 on average). ILS implements a more efficient local search than LS, and therefore during the 10 seconds, Multi-C2-ILS is able to generate and improve more

Table 2.5 Average percentage deviations.

Set	n	m	Multi-C2	Multi-C2-LS	Multi-C2-ILS
GKD−c	500	50	0.336	0.000	0.000
MDG−a	500	50	0.174	0.265	0.185
MDG−a	500	200	0.630	0.036	0.028
MDG−a	2000	200	0.447	0.334	0.232
MDG−b	500	50	0.207	0.271	0.162
SOM−b	500	200	0.297	0.056	0.037

solutions than Multi-C2-LS (around 500 solutions on average). This table shows that Multi-C2-ILS obtains better solutions than Multi-C2 in all the sets but one (MDG-a with $n = 500$ and $m = 50$), where they obtain similar results. This clearly illustrates the convenience of applying a local search post-processing to a generation method, specially if it implements candidate list strategies, as it is the case of ILS.

It should be noted that, unlike other well-known methods that we will review in the next chapter, multi-start procedures have not yet become widely implemented and tested as a meta-heuristic itself for solving complex optimization problems. We have shown new ideas that have recently emerged within the multi-start area that add a clear potential to this framework which has yet to be fully explored.

Chapter 3
Meta-Heuristics

Abstract In this chapter we elaborate on meta-heuristics for optimization from
a beginner's perspective. Basically, we start from scratch to describe the different
methodologies and provide the reader with the elements and strategies to build and
implement them successfully. Although we describe the adaptation of these methods
to the linear ordering problem (LOP) and the maximum diversity problem (MDP),
we do not limit our descriptions to these problems. Instead, we present all methods in their generic form and then adapt them to the specific case of the LOPand
the MDP. This way, the reader can easily apply these methods to a wide range of
combinatorial optimization problems.

3.1 Introduction

In the last decade a series of methods have appeared under the name of *meta-
heuristics*, which aim to obtain better results than those obtained with "traditional
heuristics". The term meta-heuristic was coined by Glover [71] in 1986. In this
book we shall use the term heuristics to refer to the classical methods as opposed
to meta-heuristics, which we shall reserve to refer to the most recent and complex
ones. In some texts one can find the expression *modern heuristics* to refer to meta-
heuristics [142]. Osman and Kelly [134] introduce the following definition:

> *"A meta-heuristic is an iterative generation process which guides a subordinate heuristic by
> combining intelligently different concepts for exploring and exploiting the search spaces
> using learning strategies to structure information in order to find near-optimal solutions."*

Meta-heuristic procedures are conceptually ranked "above" heuristics in the
sense that they guide their design. Thus, facing an optimization problem, we can
employ any of these procedures to design a specific algorithm for computing an
approximate solution.

Heuristic methods have been developed to such an extent, with new methodologies appearing "almost daily", that to offer an exhaustive survey of all of them lies

far outside of the the the scope of this monograph. Furthermore, our prime interest lies
in those that have been used successfully for the LOPand the MDP. We can list as
better established ones the following approaches:

–   *EDA. Estimation distribution algorithms*
–   *EA. Evolutionary algorithms*
–   *GA. Genetic algorithms*
–   *GRASP. Greedy randomized adaptive search procedure*
–   *GLS. Guided local search*
–   *IG. Iterated Greedy*
–   *MA. Memetic algorithms*
–   *MS. Multi-start methods*
–   *PR. Path relinking*
–   *SS. Scatter search*
–   *SA. Simulated annealing*
–   *TS. Tabu search*
–   *VNS. Variable neighborhood search*

It is important to highlight two aspects. The first one is that the above list is not
exhaustive; as we mentioned previously this would be very difficult to establish. The
second one is that some methods can be considered specializations or adaptations of
other more generic ones (a consideration that is not free from controversy). For example, "ant colony optimization" is motivated by the trails left by ants in search for
optimum paths. In our view, this could as well be interpreted as a short term memory structure in tabu search. Researchers in this field increasingly propose combinations and hybridizations among the different methods, which makes the boundary
between them less well defined, in many cases taking some ingredients from others.

Every metaheuristic procedure is built by considering that to be effective it must
include strategies for both search *diversification* and *intensification* (also known as
exploration and exploitation, respectively). Intensification refers to mechanisms and
parameter settings that encourage the addition of solution features that historically
(i.e., during the search) have been found to have merit. It also refers to strategies
that focus the search on a particular region of attraction. These strategies must be
balanced with diversification processes that expand the exploration of the solution
space. The opportunity for search diversification depends on the "optimization hori-
zon", that is the length of the optimization process (computational time).

A popular thrust of many research initiatives, and especially of publications designed to catch the public eye, is to associate various methods with processes found
in nature. The main issue with metaheuristics based on metaphors of nature is not
that finding inspiration in such processes is inherently bad. The problem is that
metaphors are cheap and easy to come by and they are often used to "window dress"
a method. Nowadays, these so-called "novel" methods employ analogies that range
from intelligent water drops, musicians playing jazz, imperialist societies, leapfrogs,
kangaroos, all types of swarms and insects and even mine blast processes. In the serious metaheuristic literature, researchers that use metaphors to stimulate their own
ideas about new methods must translate these ideas into metaphor-free language, so

that the strategies employed can be clearly understood, and their novelty is made clearly visible.

All meta-heuristic methodologies have many degrees of freedom and the user must take several decisions in order to design the final algorithm. Hertz and Widmer [85] identified two categories on which meta-heuristics are based: *local search* and *population search*. In the former the method iterates over a single solution, while in the later the method iterates over a set of solutions. TS, SA or VNS are local search based methods, while SS or GA are population search based methods (in which a set of solutions evolves during the search process). In their study the authors provide the following guidelines to help designing a good method for a given problem within each category.

Local search

– It should be easy to generate solutions in the search space under consideration.
– The solutions in the neighborhood of a solution should be close to this solution in some way.
– The topology induced by the objective function in the neighborhood space should not be too "flat".
– Each solution in the search space should be linked with the optimum solution by a (short) sequence of moves.

Population search

– Pertinent information should be transmitted during the combination phase (in which new solutions are obtained with the combination of old ones).
– The combination of two equivalent *parent solutions* should not produce a new solution that is different from the parents.
– Diversity should be preserved in the population.

In this chapter we are going to describe the application of the main meta-heuristic procedures to the LOPand the MDP. We will consider only those procedures that are relatively consolidated and that have proved efficient for a representative collection of problems. Specifically, we shall describe the implementations of GRASP, IG, TS, SA, VNS, SS, and GA.

3.2 GRASP

The GRASP methodology was developed in the late 1980s, and this acronym for *greedy randomized adaptive search procedures* was coined by Feo and Resende [57]. It was first used in [56] to solve computationally difficult set covering problems. Each GRASP iteration consists of constructing a trial solution and then applying an improvement procedure to find a local optimum (i.e., the final solution for that iteration). The construction phase is *iterative, greedy* and *adaptive*. It is iterative because the initial solution is built considering one element at a time. It is greedy because the addition of each element is guided by a greedy function. It is adaptive because the element chosen at any iteration in a construction is a function of those chosen previously and thus relevant information is updated from one construction step to the next. The improvement phase typically entails a local search procedure. The GRASP method is similar in layout to the general multi-start methods described in the previous chapter.

GRASP

(1) Set $i=1$.

(2) While $i < MaxIter$.

    (2.1) Construct a solution x_i. (*Construction phase*)

    (2.2) Apply a local search method to improve x_i. Let x_i' be the solution obtained. (*Improvement phase*)

    (2.3) If x_i' improves the best solution found, update it. (*Test*)

    (2.4) $i = i+1$.

We now describe the two phases in more detail.

3.2.1 Construction Phase

At each iteration of the construction phase, GRASP maintains a candidate list CL of elements which can be added to the partial solution under construction to obtain a feasible complete solution. All candidate elements are evaluated according to a greedy function in order to select the next element to be added in the construction. The greedy function usually represents the marginal increase in the cost function by adding the element to the partial solution. Element evaluation is used to create a restricted candidate list RCL with the best elements in CL, i.e., those with the largest incremental cost in a maximization problem (as in the LOPand MDP). The element to be added to the partial solution is randomly selected from those in the RCL. Once the selected element is added to the partial solution, the candidate list is updated and the evaluations (incremental costs) are recalculated.

A particularly appealing characteristic of GRASP is its ease of implementation. We only need to define a construction mechanism according to the description above and a local search procedure. Moreover, the construction usually has one parameter related to the quality of the elements in the restricted candidate list. GRASP typically performs a pre-established number of iterations (construction + improvement) and returns the best solution found overall.

The Linear Ordering Problem

The first step in designing a GRASP construction is to define a greedy function in order to evaluate the relative contribution of adding an element to the partial solution under construction. In [25] the following three evaluators are proposed for the LOP, inspired by the simple heuristics described in the previous chapter:

- The evaluation $e_1(i)$ of object i is the sum of the elements in the matrix in its corresponding row:

$$e_1(i) = \sum_{j=1}^{n} c_{ij}.$$

- The evaluation $e_2(i)$ of object i is the difference between the maximum of the sum of columns, *colmax*, and the sum of the elements in the matrix in its corresponding column:

$$e_2(i) = colmax - \sum_{j=1}^{n} c_{ji}.$$

- The evaluation $e_3(i)$ of object i is the sum of the elements in the matrix in its corresponding row divided by the sum of the elements in its corresponding column:

$$e_3(i) = \frac{\sum_{j=1}^{n} c_{ij}}{\sum_{j=1}^{n} c_{ji}}.$$

The goal of the constructive method is to obtain a permutation of size n, so in each iteration we select an object from an RCL to be placed in the next available position. The permutations are constructed from left to right, meaning that we start from position 1 and end at position n. Initially, all objects are in the unassigned list U. For each object $i \in U$, we calculate its associated evaluation $e_1(i)$ (or alternatively $e_2(i)$ or $e_3(i)$) with respect to the unassigned objects (i.e., we only compute in the expressions above the c-values associated to unassigned objects). This measures the attractiveness of each object. The largest attractiveness value e_1^* of all the unassigned objects is multiplied by a parameter α, $0 \leq \alpha \leq 1$. This value represents a threshold that is used to build the RCL. In particular, the RCL comprises all the objects in U with an attractiveness measure that is at least as large as the threshold value and we set RCL $= \{i \in U \mid e_1(i) \geq \alpha e_1^*\}$, where $e_1^* = \max\{e_1(i) \mid i \in U\}$.

Then, the next object to be assigned is randomly selected from the RCL. After the assignment has been made, the list of unassigned objects is updated and the measure of attractiveness for the objects in the updated set U is recalculated. The process is repeated n times and the outcome is a feasible solution of the LOP.

The example in Figure 3.1 shows the value of 100 solutions generated with the construction described above. We apply the evaluator e_1 and create the RCL with α set to 0.4 to solve the LOLIB instance be75eec. The best solution found in these 100 iterations has value 199072. Note that GRASP performs an independent sampling of the solution space, and therefore there is no connection between different constructions. In other words, we cannot predict the value of a solution by considering the value of the previously generated solutions. This is why we cannot find any pattern in Figure 3.1.

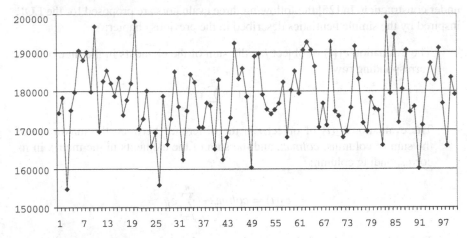

Fig. 3.1 Value of GRASP constructions

The design of different evaluation functions for the GRASP construction phase leads us to an interesting point: How can we compare different constructive methods? This is not a specific question about GRASP, but it is a general question about constructive methods. Obviously we want to obtain good solutions, so the larger the value of the solutions obtained (in a maximization problem), the better the constructive method. However, as discussed in the previous chapter in the context of general multi-start methods, we need to establish a trade-off between quality and diversity when selecting our construction method.

Let GC_1, GC_2 and GC_3 be the three GRASP constructive methods obtained with the greedy functions e_1, e_2 and e_3 respectively. In order to compare their relative performance when solving the `LOLIB` instance `be75eec`, we generate 100 solutions with each method and then compute a measure of both the quality and diversity of all 100 solutions. To evaluate quality we can simply compute the average objective function value of the 100 solutions generated with each method. To evaluate diversity we can compute the distance between each pair of solutions and then calculate the average of these distances. The distance between two orderings or permutations $p = \langle p_1, p_2, ..., p_n \rangle$ and $q = \langle q_1, q_2, ..., q_n \rangle$ is given by

$$d(p,q) = \sum_{i=1}^{n} |p_i - q_i|.$$

Table 3.1 shows both measures for the three constructive methods. We can see that GC_3 provides the best results in terms of quality, since it obtains an average objective function value of 222013.87, which compares favorably with 178080.9

obtained with GC_1 and 161885.1 obtained with GC_2. On the other hand, GC_2 provides the best results in terms of diversity, since it obtains an average distance value of 802.5, which compares favorably with 743.6 obtained with GC_1 and 632.9 obtained with GC_3. However, if we consider both criteria together, quality and diversity, GC_1 is probably the best constructive method, providing a good trade-off between them.

Table 3.1 Comparison of constructive methods

Method	Quality	Diversity
GC_1	178080.90	743.6
GC_2	161885.10	802.5
GC_3	222013.87	632.9

   GRASP may be viewed as a repetitive sampling technique, producing a sample solution in each iteration from an unknown distribution, whose mean and variance are functions of the restrictive nature of the RCL. For example, if RCL is restricted to a single element, then only one solution will be produced, the variance of the distribution will be zero and the mean will equal the value of the greedy solution. If the RCL has more than one element, then many different solutions will be generated, implying a larger variance. Since greediness plays a less important role in this case, the mean solution value may be worse. However, the value of the best solution found outperforms the mean value and can be better than the greedy solution (and in some cases may be optimal).

   In the definition of the restricted candidate list RCL given above, we can see that only the elements with a value exceeding αe_1^* are included in this list. We can therefore interpret the parameter α as the degree of greediness (between 0 and 1) in the selection of the element to be added to the partial solution. If α takes the value 0, all the candidate elements will be included in RCL and the method will actually perform a random selection (not greedy at all). On the contrary, if α takes the value 1, only the most highly evaluated elements will be included in RCL, and the method will perform a greedy selection (with no randomization at all). It is then expected that low values of α will produce diverse solutions of low quality, while large values of α will produce similar solutions (low diversity) of good quality.

   To test this hypothesis, we consider the LOLIB instance be75eec and generate 100 solutions with the constructive method GC_1 with each particular value of α in the set {0.0, 0.1, 0.2, ..., 1.0}. Figure 3.2 shows the average value of the quality and diversity of each set of 100 solutions, obtained as in the previous experiment, for each value of α. In order to compare both values, quality and diversity, we represent them as a percentage and then we add them up to obtain a global evaluation of GC_1 with each α value considered.

   Figure 3.2 shows that, as expected, when α is set to 0, the average quality obtained is the lowest (0% in the figure) but the diversity is the largest (100% in the figure). Symmetrically, when α is set to 1, the average quality obtained is close to the largest one (95% in the figure) but the diversity is the lowest one (0% in the figure). If we consider the addition of both percentage values (which means that we consider quality and diversity as equally important), we can see that when α is set to 0.4 we obtain the best solutions (185% in the figure). Therefore we will use GC_1 with α set to 0.4 as the constructive method in our GRASP algorithm for the LOP.

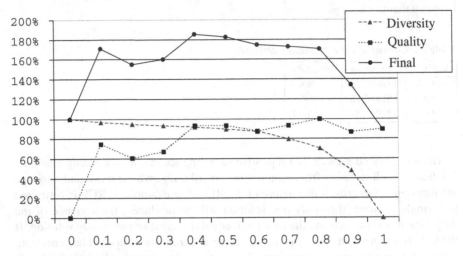

Fig. 3.2 Performance according to α

In [38] an alternative GRASP construction is proposed. It is based on a variant of the evaluation function $e_3(i)$ in which, instead of ratios, differences are computed. This way, the evaluation $e(i)$ of object i is the sum of the elements in the matrix in its corresponding row minus the sum of the elements in its corresponding column, i.e., $e(i) = \sum_{j=1}^{n} c_{ij} - \sum_{j=1}^{n} c_{ji}$.

The authors compared the use of $e(i)$ with $e_3(i)$ and did not observe significant differences in the quality of solutions, although they stated that adapting the greedy function is computationally more expensive when ratios are used.

The construction method in [38] follows the same steps as the other GRASP methods described in this section (constructing first a candidate list of elements and computing then a restricted list RCL from which random selections are made). However, instead of inserting the selected element in the next available position, it inserts the element in the best position according to the value of the elements already selected. This strategy, introduced in [31], can be considered a local search in a reduced (partial) neighborhood.

The Maximum Diversity Problem

Silva et al. [153] proposed in 2004 several heuristics based on the GRASP methodology for the MDP. They combined different constructions with local searches and tested them on a wide set of instances, which includes the largest reported so far at that time. They called KLD to the basic construction algorithm, and KLDv2 to the improved version. It must be noted that in these largest instances with $n = 500$ elements, the methods require many hours of running time.

In 2007, Duarte and Martí [51] applied two metaheuristics for the MDP. Specifically, the authors introduced a tabu search and two GRASP algorithms, called GRASP-C2 and GRASP-D2, based respectively on the constructive method C2, and the destructive method, D2, described in the previous chapter. To explain the construction phase in GRASP-C2, we consider a graph $G_n = (V_n, E_n)$ with node set $V_n = \{1, 2, \ldots, n\}$, and where d_{ij} is the distance between nodes i and j.

The algorithm first populates the partial solution under construction M with the two elements, s and t, at the largest distance. M is initialized with these elements ($M = \{s, t\}$ and $f(M) = d_{st}$) in step 3 of the pseudo-code, and the candidate set of elements, CL, which contains all non-selected elements ($CL = V_n \setminus M$) is created in step 4. The initialization of the method finishes in step 5, in which the contribution of all the candidate elements is established as the sum of the distances to the two elements in M.

GRASP Construction for the MDP

(1) Let $G_n = (V_n, E_n)$ be a graph with distance matrix $D = (d_{ij})$

(2) $d_{st} \leftarrow \max\limits_{(i,j) \in E_n} d_{ij}$

(3) $M \leftarrow \{s, t\}$ and $f(M) = d_{st}$

(4) $CL = V_n \setminus M$

(5) $cont(j) = d_{js} + d_{jt} \; \forall j \in CL$

(6) WHILE $|M| < m$

    (6.1) $cont_{max} \leftarrow \max\limits_{j \in CL} cont(j)$

    (6.2) $cont_{min} \leftarrow \min\limits_{j \in CL} cont(j)$

    (6.3) $RCL \leftarrow \{j \in CL \mid cont(j) \geq cont_{max} - \alpha \cdot (cont_{max} - cont_{min})\}$

    (6.4) $i \leftarrow \texttt{SelectRandom}(RCL)$

    (6.5) $M \leftarrow M \cup \{i\}$ and $f(M) = f(M) + cont(i)$

    (6.6) $CL \leftarrow CL \setminus \{i\}$

    (6.7) $cont(i) = 0$

    (6.8) $cont(j) = cont(j) + d_{ij} \; \forall j \in CL$

To create the restricted candidate list, RCL, with the good elements that could be added to M, we first calculate the maximum, $cont_{max}$, and the minimum, $cont_{min}$, of the contributions in steps 6.1 and 6.2 respectively. The contribution of an element j, $cont(j)$, computes the sum of the distances between j and the elements in M. In a construction method, it basically reflects the potential addition to the objective function if the element is included in the partial solution under construction. In step 6.3, RCL is calculated as the set of candidate elements with a contribution within a percentage α of this maximum. In mathematical terms:

$$RCL = \{ j \in CL \mid cont(j) \geq cont_{max} - \alpha \cdot (cont_{max} - cont_{min}) \} \qquad (3.1)$$

In step 6.4 of the pseudo-code above, an element i is randomly selected from RCL. Note that if $\alpha = 0$ in Equation 3.1, RCL only contains the elements with a contribution larger than or equal to the maximum, which means equal to the maximum (since a value cannot be larger than the maximum value). A selection in that RCL is equivalent to a greedy selection of the best candidate. On the other hand, if $\alpha = 1$ in Equation 3.1, we obtain a RCL with those elements with a contribution larger than or equal to the minimum contribution, which means all the candidate elements (since all of them meet this condition). A selection in that RCL is equivalent to a completely random method, in which we add a randomly selected element to the partial solution. In this way, we can say that $\alpha = 0$ is a greedy method, and $\alpha = 1$ is a random method. Values of α in [0,1] would therefore reflect a combination of greediness and randomization.

The core of the GRASP algorithm takes place in the while loop from steps 6.1 to 6.8, reflecting the successive applications of this part of the code. In each one, the RCL is created, an element $i \in RCL$ is randomly selected, and the different elements of the algorithm are updated. In particular, in step 6.5, i is included in M and its value updated $(f(M) = f(M) + cont(i))$, and in step 6.6, it is removed from CL. Finally, the contribution value of all the non-selected elements is increased in step 6.8 by adding their distance to i. Note that to prevent the selection of already selected elements, we set the contribution of i to 0 in step 6.7. The method finishes when the while loop performs $m - 2$ iterations (i.e., when $|M| = m$).

The description above on GRASP-C2 can be easily adapted to obtain GRASP-D2. Starting with all the elements selected, GRASP-D2 basically deselects elements, one by one, until only m of them remain selected. The restricted candidate list contains now the elements with low contribution, and the method selects one element at random from it at each iteration. The element is removed (deselected) from the partial unfeasible solution, and the method updates the contribution, similarly to GRASP-C2.

3.2.2 Improvement Phase

The second stage of a GRASP algorithm consists in improving the constructed solutions using a local search method, which guides the search process to a local optimum. Local search methods are based on the notion of neighbor structures that generate changes to move from one solution to another in the solution space. Local searches perform moves as long as the current solution improves and terminate when no further improvement is possible. The resulting solution is said to be locally optimal (i.e., the solution cannot be improved within the defined neighbor structure).

In this section we apply a local search to improve the solution of the LOP and the MDP. Local search is defined in terms of the solution representation, and as explained in detail in previous sections, we represent LOP solutions with permutations, and MDP solutions with binary vectors. As stated in Duarte et al. [52], how to represent a solution is a fundamental design question in metaheuristic optimization. The solution representation is the main factor that determines the form and size of the search space. The representation also determines how the objective function will be evaluated and how the move mechanisms (also referred to as search operators) will create neighborhoods. In this section we consider an insertion move for the LOP and an exchange move for the MDP.

The Linear Ordering Problem

The local search of our GRASP algorithm is based on the neighborhood search developed for the LOP presented in [106]. Here insertions are used as the primary mechanism to move from one solution to another. We again define $move(p_j, i)$ as the local modification deleting p_j from its current position j in permutation p and inserting it at position i (i.e., between the objects currently at positions $i-1$ and i).

A key to designing an efficient local search procedure is the incremental computation of the solution value. In other words, when we move from one solution p to another q, we need to calculate the objective function value of q from the value of p quickly and efficiently. In the LOP, when we apply $move(p_j, i)$ to solution p, obtaining solution q, it is easy to check that the value $z(q)$ of q can be directly obtained from the value $z(p)$ by considering only the elements in row and column j in the data matrix C. More precisely, if $i < j$ then

$$z(q) = z(p) + \sum_{k=i}^{j-1} (c_{p_j p_k} - c_{p_k p_j}).$$

Starting from a solution generated in the construction phase, the improvement phase of our GRASP algorithm performs iteration steps as long as the objective function increases. At each step, it considers an element p_j in p and scans all positions i from 1 to n, in search for the position i^* with the best associated move $move(p_j, i^*)$. If this is an improving move, it is performed; otherwise it is discarded and the method resorts to the next element p_{j+1} in the solution.

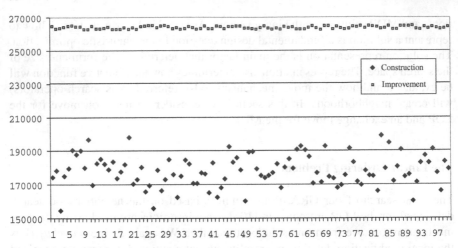

Fig. 3.3 Construction and local search values

Figure 3.3 shows the value of 100 solutions constructed with GC_1 (with α set to 0.4) and the value of the improved solutions with this local search procedure on the LOLIB instance be75eec. The effectiveness of the local search procedure is clearly shown, since it is able to improve the solutions by 48.5% on average. It also shows a large variability in the objective function of the constructed solutions, with a minimum value of 154728 and a maximum of 199072. However, on the contrary, it shows very similar values in the improved solution, with a minimum of 262387 and a maximum of 264940. Therefore, we cannot conclude that constructed solutions with low quality produce low quality improved solutions. This would indicate that diversity is more important than quality in the constructive method.

In our final experiment in this section, we apply the GRASP algorithm for 10 seconds to the 229 instances in the OPT-I set (with optimum known). Table 3.2 shows the number of instances in each set, the average percentage relative deviation *Dev* between the best solution value found with this method and the optimal value, and the number of instances, *#Opt*, for which an optimum solution is found.

Table 3.2 GRASP on OPT-I instances

	IO	SGB	RandAII	Rand B	MB	Special	**Total**
#Instances	50	25	25	70	30	29	229
Dev(%)	0.00	0.00	0.01	0.00	0.00	0.05	0.01
#Opt	49	23	5	70	21	14	182

The results in Table 3.2 clearly indicate that OPT-I instances do not pose a challenge for the GRASP methodology, since it is able to obtain 182 optimum solutions out of 229 instances in 10 seconds of running time. We will see in the next sections that most of the meta-heuristic are able to obtain high quality solutions on these instances within short running times.

An interesting analysis of the local search method is the study of the objective function values of the solutions in the neighborhood of a local optimum. This is usually called *landscape analysis*, and may disclose important information that can improve the exploration. Hernando et al. [84] consider three typical moves for the LOP: insertions, adjacent swaps, and general swaps. The authors focus on the number and the identification of solutions that may be discarded from exploration, when a local optimum is known, because they will not improve that optimum. To illustrate the impact of their finding, it is shown that in an instance with $n = 20$ elements, in which we apply a local search based on insertions, around 10^{10} solutions can be discarded from the exploration, with the corresponding computational time save.

The Maximum Diversity Problem

Duarte and Martí [51] proposed several strategies to explore the typical neighborhood of the MDP based on exchanges in an efficient way, to avoid the long running times of previous tabu search and GRASP implementations. In particular, instead of searching for the best exchange at each iteration, their neighborhood exploration performs two stages. In the first one, it selects the element with the lowest contribution to the value of the current solution. Then, in the second stage, the method performs the first improving move to replace it (i.e., instead of scanning the whole set of unselected elements searching for the best exchange, it performs the first improving exchange without examining the remaining unselected elements). Their experimentation confirms the effectiveness of the proposed strategies.

One of the key elements in designing an effective local search method is the definition of the *move* and the associated move value (change in the objective function value). In particular, for the MDP we define $\text{move}(M, i, j)$ as the move that interchanges vertex $i \in M$ with vertex $j \in V \setminus M$. This move usually produces a variation in the objective function, denoted as $\text{move\_value}(M, i, j)$. If the move value is positive, it indicates that in a maximization problem, such as the MDP, the associated move improves the solution. We therefore would apply the move and obtain a new solution better than M.

Given a solution M, an element $i \in M$, and an element $j \in V \setminus M$, the move value can be computed as follows. If i is removed from the solution, its value would decrease in the sum of the distances from i to the rest of the elements in M. In mathematical terms,

$$out\_value(i) = \sum_{u \in M} d_{ui}.$$

On the other hand, if j is added to the solution, its value would increase in the sum of the distances from j to the elements in M (excluding the i). In mathematical terms,

$$in\_value(j) = \sum_{u \in M, u \neq i} d_{uj}.$$

Then, the move value can be computed from these two values as:

$$move\_value(M, i, j) = in\_value(j) - out\_value(i)$$

The local search method scans, at each iteration, the list of elements in the solution ($i \in M$) in lexicographical order (i.e., from the first to the last one). The pseudo-code below shows a *FOR* statement in line 3, in which the index i takes values from 1 to m to examine M in a loop. Then, for each element, $i \in M$, it examines the list of unselected elements ($j \in V \setminus M$) in search for the first improving exchange (i.e., $move\_value(M, i, j) > 0$). The unselected elements are also examined in lexicographical order (from 1 to n) in line 3.3.1, where the elements are selected from CL.

Line 3.3.3 in the pseudo-code checks if the move improves the solution; and in that case, the method performs the first improving move. To do so, it first updates the set M with:

$$M \leftarrow M \setminus \{i\} \cup \{j\}$$

It then updates the objective function value $f(M)$, by adding the move value to it.

$$f(M) \leftarrow f(M) + in\_value(j) - out\_value(i)$$

Finally, the candidate set of elements CL is updated with the following expression:

$$CL \leftarrow CL \setminus \{j\} \cup \{i\}$$

This concludes the current iteration in the inner *WHILE* loop (line 3.3). The algorithm repeats iterations as long as improving moves can be performed, and stops when no further improvement is possible. Note that when an improving move is performed, the i index is reset to 0, and then the *FOR* loop is initiated again in 1. In this way, the method only terminates when all the elements in the solution have been examined an no move has been performed (step 3.4). We can then assure that the current solution is a local optimum.

Local search for the MDP

(1) Let M be a solution with m elements, and $f(M)$ its value
(2) $CL = V \setminus M$
(3) FOR $i = 1$ to m

   (3.1) Compute $out\_value(i)$
   (3.2) Move=0
   (3.3) WHILE $Move = 0$ and CL not explored
     (3.3.1) $j \leftarrow$ SelectNext(CL)
     (3.3.2) Compute $in\_value(j)$
     (3.3.3) IF $in\_value(j) > out\_value(i)$
        Perform the move and make $Move = 1$ and $i = 0$
   (3.4) If no move has been performed: STOP.

3.3 Strategic Oscillation and Iterated Greedy

Strategic oscillation (SO) [69] operates by orienting moves in relation to a critical level, which in our problem is defined as the number of selected elements m. Such a critical level or oscillation boundary often represents a point where the method would normally stop. Instead of stopping when this boundary is reached, the rules for selecting elements are modified, to permit the region defined by the critical level to be crossed. The approach then proceeds for a specified depth beyond the oscillation boundary, and turns around. The oscillation boundary again is approached and crossed, this time from the opposite direction, and the method proceeds to a new turning point. The process of repeatedly approaching and crossing the critical level from different directions creates an oscillatory behavior, which gives the method its name. This dynamic neighborhood approach applies not only to the types of neighborhoods used in solution improvement methods (usually called -local search methods-) but also applies to constructive neighborhoods used in building solutions from scratch - as opposed to transitioning from one solution to another.

The application of the SO framework usually exploits problem constraints as a source of critical boundaries that the method may explore. A typical implementation iteratively applies three stages:

1. An oscillation process explores the feasible and infeasible regions around a current solution and returns a new candidate.
2. A local optimization operator is applied to every new candidate to get an associated improved solution from the nearby area of the search space.
3. An acceptance criterion decides which improved solution is chosen to continue the search. Additionally, we may employ a constructor operator to build initial solutions at the beginning of the run and when stagnation is detected.

The acceptance-criterion of establishes the rules by which the algorithm wanders over the regions of the search space in the quest for better solutions. The following two criteria are the standard from the literature:

- Replace-if-better: The new solution is accepted only if it attains a better objective function value.
- Random walk: Several authors have pointed out that the foregoing acceptance criterion may lead the method to stagnation, because of getting trapped in a local optimum. By contrast, the random walk criterion always selects the new solution, regardless of its objective function value, which prevents the method from being confined in the area of one local optimum.

More recently, constructive and destructive neighborhoods have been applied within a simplified and effective method known as iterated greedy (IG) [150], which generates a sequence of solutions by iterating over a greedy constructive heuristic which, like strategic oscillation, uses two main phases: destruction and construction. IG is a memoryless metaheuristic easy to implement that has exhibited state-of-the-art performance in some settings (see for example [112]).

Glover et al. [68] proposed two constructive methods, C1 and C2, and two destructive methods, D1 and D2 for the Maximum Diversity Problem. Constructive method C1 adds elements, one-by-one, to the current solution under construction, up to m elements have been added. At this point it has a feasible solution, say M_1 and the method stops. The SO method basically consists in "keep going" a few more steps. In particular, it considers the addition of extra elements to the solution M_1, with the same method C1, obtaining an unfeasible solution M_2. Then, it will apply a destructive method, such as D1, to remove from M_2 some elements. In this way, we may obtain a new feasible solution, M_3, that eventually could be better than the two previous ones.

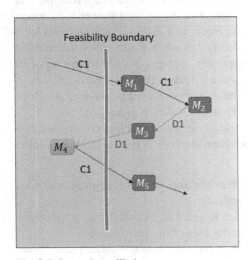

Fig. 3.4 Strategic oscillation

Figure 3.4 illustrates the strategic oscillation. In line with the "keep-going" strategy described above when adding elements, it also proposes to remove a few additional elements from the feasible solution. In particular, we may apply D1 to remove some extra elements from M_3, thus obtaining a partial, unfeasible, solution M_4. If we repeat this scheme, adding now elements to M_4 with C1, obtaining a new feasible solution M_5, we have an oscillation pattern, crossing the feasibility boundary of the solution space.

Further studies are required to evaluate the efficiency of this methodology, compared with the GRASP described in previous sections. We anticipate an important area of research when comparing both methodologies, since each one somehow represents a class of methods. In GRASP we perform independent constructions, based on the random sampling schema. On the contrary, in IG we construct a solution from the partial solution obtained destructing the previous one, and therefore two consecutive feasible solutions are linked (dependent). This method is based on the idea of taking advantage of information previously recorded in the search, that is the key-point of the tabu search methodology that we describe now.

3.4 Tabu Search

Tabu search [69] is a meta-heuristic that guides a local heuristic search procedure to explore the solution space beyond local optimality by allowing non-improving moves. This, in turn, requires some additional mechanisms based on memory structures to avoid cycling, which creates flexible search behavior. The name *tabu* comes from the fact that move selection is limited because certain moves or solutions are forbidden (declared tabu) as a result of the recorded information. In this section we will see how to implement both simple and complex memory structures that allow us to record useful information (called *attributes*) about the solutions visited during the search. Based on this information we will design a local search method with attributive and adaptive memory.

Tabu search (TS) uses *attributive memory* for guiding purposes. This type of memory records information about solution attributes that change in moving from one solution to another. For example, in a graph or network setting, attributes can consist of nodes or arcs that are added, dropped or repositioned by the moving mechanism. In production scheduling, the index of jobs may be used as attributes to inhibit or encourage the method to follow certain search directions.

We can compare the memory of tabu search with the *explicit memory* of branch-and-bound methods (see Chap. 4) in which an exhaustive memory is applied to know the solutions we have already examined during the search process. In contrast, TS incorporates *attributive* and *adaptive* memory. The attributive term means that we are not going to record all the solutions generated (because we are not interested in examining all the solutions of the problem) but we are going to record some of their properties (attributes). The term "adaptive" means that the rules and properties that we are going to apply may evolve during the search process.

In this section we start describing the so-called *short term memory* design. In this design we record attributes of recently visited solutions to modify the neighborhood with them. Then, we will see how we can add *long term memory* strategies to this basic design to obtain a competitive solution method. Both memory strategies together, short and long term, constitute the core of most tabu search implementations.

If we see the search space as a huge set of solutions and think that we are only able to explore a tiny part of it, we easily understand the rationale behind the two search strategies *intensification* and *diversification*. Roughly speaking, the first one favors the exploration of promising areas of the solution space, while the second one drives the search to new regions of the solution space. Therefore, intensification and diversification have complementary objectives. Intensification strategies are based on modifying choice rules to encourage move combinations and solution features that have historically been found to be good. In some settings, they consist of revisiting attractive regions to search them more thoroughly. Diversification strategies, on the other hand, are based on visiting unexplored regions, increasing the effectiveness in exploring the solution space of search methods based on local optimization. Both strategies interact with the two memory structures mentioned above, short and long term, to create an efficient search algorithm.

3.4.1 Short Term Memory

Tabu search begins in the same way as an ordinary local or neighborhood search, proceeding iteratively from one solution to another until a chosen termination criterion is satisfied. When TS is applied to an optimization problem with the objective of minimizing or maximizing $f(x)$ subject to $x \in X$, each solution x has an associated neighborhood $\mathcal{N}(x)$, and each solution $y \in \mathcal{N}(x)$ can be reached from x by an operation called a *move*.

When contrasting TS with a simple descent method where the goal is to minimize $f(x)$, we must point out that such a method only permits moves to neighbor solutions that improve the current objective function value and ends when no improving solutions can be found. Tabu search, on the other hand, permits moves that deteriorate the current objective function value but the moves are chosen from a modified neighborhood $\mathcal{N}^*(x)$. Short and long term memory structures are responsible for the specific composition of $\mathcal{N}^*(x)$. In other words, the modified neighborhood is the result of maintaining a selective history of the states encountered during the search.

In short term memory we usually consider attributes of the solutions recently visited, or the moves performed, in the last iterations. These attributes are used to exclude some elements in the neighborhood of the solutions in the next iterations. $\mathcal{N}^*(x)$ is typically a subset of $\mathcal{N}(x)$, and the *tabu classification* serves to identify elements of $\mathcal{N}(x)$ excluded from $\mathcal{N}^*(x)$. Let $\mathcal{T}(x)$ be this set of solutions in $\mathcal{N}(x)$, labeled as tabu, that we do not consider admissible for selection, i.e.,

$$\mathscr{N}^*(x) = \mathscr{N}(x) \setminus \mathscr{T}(x).$$

Recency based memory, as its name suggests, keeps track of attributes of solutions that have changed during the recent past. To exploit this memory, selected attributes that occur in solutions recently visited are labeled *tabu-active*, and solutions that contain tabu-active elements, or particular combinations of these attributes, are those that become tabu, thus being included in $\mathscr{T}(x)$. This prevents certain solutions from the recent past from belonging to $\mathscr{N}^*(x)$ and hence from being revisited.

The Linear Ordering Problem

The basic tabu search algorithm for the LOP proposed in [106] implements a short term memory structure alternating two phases: intensification and diversification. It is based on insertions as the improvement phase of the GRASP algorithm described above. However, instead of scanning the objects in the search for a move in their original order (from 1 to n), they are randomly selected in the intensification phase based on a measure of influence.

Short term tabu search

(1) Generate an initial solution x.
(2) Determine the attributes to establish the tabu status of solutions.
(3) While the stopping condition is not satisfied:

    (3.1) Compute $\mathscr{T}(x)$ and set $\mathscr{N}^*(x) = \mathscr{N}(x) \setminus \mathscr{T}(x)$.
    (3.2) Let y be the best solution in $\mathscr{N}^*(x)$.
    (3.3) Set $x = y$.
    (3.4) Update the tabu status of solutions.

For each object, there are at most m, $m \leq 2n - 2$, relevant elements in the matrix, i.e., those elements that may contribute to the objective function value. The elements in the main diagonal of the matrix are excluded because their sum does not depend on the ordering of the objects. This indicates that objects should not be treated equally by a procedure that selects an object for a local search (i.e., for search intensification). We define w_j as the weight of object j by setting

$$w_j = \sum_{i \neq j} (c_{ij} + c_{ji}).$$

Note that weight values do not depend on the permutation p, and therefore they can be calculated off-line, i.e., before the search begins. The weight values will be used to bias the selection of objects during the tabu search intensification phase.

An iteration in the *intensification phase* begins by randomly selecting an object. The probability of selecting object j is proportional to its weight w_j. The move $move(p_j, i)$ with the largest move value is selected. (Note that this rule may result in the selection of a non-improving move.) The move is executed even when the move value is not positive, resulting in a deterioration of the current objective function value. The moved object becomes tabu-active for *TabuTenure* iterations, and therefore it cannot be selected for insertions during this time.

The number of times that object j has been chosen to be moved is accumulated in the value *freq*(j). This frequency information is used for diversification purposes. The intensification phase terminates after *MaxIter* consecutive iterations without improvement.

The *diversification phase* is performed for *MaxDiv* iterations. In each iteration, an object is randomly selected, where the probability of selecting object j is inversely proportional to the frequency count *freq*(j). The chosen object is placed in the best position, as determined by the move values associated with the insert moves.

The basic tabu search procedure stops when *MaxGlo* global iterations are performed without improving the value of the best solution p^* found. A global iteration is an application of the intensification phase followed by the application of the diversification phase.

The Maximum Diversity Problem

The short term tabu search proposed by Duarte and Martí [51] for the MDP, LS-TS, is based on exchanges, as the local search for the MDP described in the GRASP methodology. In particular, it also implements the move move(M, i, j) that interchanges vertex $i \in M$ with vertex $j \in V \setminus M$.

An important characteristic of this method is that, instead of exploring the neighborhood following the canonical order, it computes the sum of distances, $d(i)$, from each element i in the solution to the rest of the elements, and explores first the solutions in the neighborhood associated to remove from the solution the elements with a lower contribution to the objective function (i.e., sum of distances). In mathematical terms:

$$d(i) = \sum_{u \in M} d_{ui}.$$

An iteration in this method begins by randomly selecting an element i in M. The probability of selecting element i is inversely proportional to its $d(i)$-value. The first improving move move(M, i, j) associated with i is performed. (Note that if there is no improving move associated with i, we perform the best one available, even if it is a non-improving move.) The move is executed even when the move value, move_value(M, i, j), is not positive, resulting in a deterioration of the current objective function value. The moved elements i, j become tabu-active for TabuTenure iterations, and therefore they cannot be unselected (respectively selected) during this time. The LS-TS method performs iterations until in MaxIter consecutive iterations the best solution has not been improved.

3.4.2 Long Term Memory

In TS strategies based on short term memory, $\mathcal{N}^*(x)$ characteristically is a subset of $\mathcal{N}(x)$, and the tabu classification serves to identify elements of $\mathcal{N}(x)$ excluded from $\mathcal{N}^*(x)$. In TS strategies that include longer term considerations, $\mathcal{N}^*(x)$ may also be expanded to include solutions not ordinarily found in $\mathcal{N}(x)$, such as solutions visited and evaluated in a past search, or identified as high quality neighbors of these past solutions.

The most common attributive memory approaches are *recency* based memory and *frequency* based memory. Recency, as its name suggests, keeps track of solutions' attributes that have changed during the recent past and we usually apply it in short term memory as described in the previous section. Frequency based memory provides a type of information that complements the information provided by recency based memory, broadening the foundation for selecting preferred moves. Like recency, frequency often is weighted or decomposed into subclasses. Also, frequency can be integrated with recency to provide a composite structure for creating penalties and inducements that modify move evaluations.

Frequencies typically consist of ratios, whose numerators represent the number of iterations (that we will refer to as *transition measures*) where an attribute of the solutions visited changes, and the denominators generally represent the total number of associated iterations. Alternatively, the numerators can represent the number of iterations where an attribute belongs to solutions visited on a particular trajectory (*residence measures*). Therefore, the ratios can produce transition frequencies that keep track of how often attributes change, or residence frequencies that keep track of how often attributes are members of solutions generated.

The Linear Ordering Problem

A long term diversification phase to complement the basic tabu search algorithm for the LOP is implemented in [106]. The long term diversification is applied after a pre-established number of global iterations have elapsed without improving the value of p^*.

For each object p_j, a rounded average position $\alpha(p_j)$ is calculated using the positions occupied by this object in the set of elite solutions and the solutions visited during the last intensification phase. The set of elite solutions consists of the best solutions found during the entire search. Instead of only recording (and updating) the best solution p^* found during the search, the long term memory records (and updates) a set of best solutions. Then, it calculates the average position occupied by each object in this set.

As mentioned above, in the computation of $\alpha(p_j)$ we also include the position occupied by the object during the last application of the intensification phase. Roughly speaking, we want to obtain a new solution far from those in the elite set but also far from those visited in the last iterations.

The long term diversification phase performs n steps, scanning the objects in their original order for $j = 1$ to n. In step j we insert object p_j in its complementary position $n - \alpha(p_j)$ according to the average position computed above. In mathematical terms, we apply $move(p_j, n - \alpha(p_j))$.

This strategy is inspired by the REVERSE operation of [31]. We, however, incorporate information about solutions that have been recently visited (during the last intensification phase) and solutions of high quality that have been found during the search (elite solutions). Purposefully constructing solutions that are "far away" from those in the elite set constitutes a diversifying element that also complements the diversification described in the short term memory.

In [106] series of experiments are performed to first establish the value of the key search parameters, and then to compare their method with the state of the art procedures. The first experiment has the goal of finding appropriate values for the three critical search parameters: *TabuTenure*, *MaxIter*, and *MaxDiv*. With a full factorial design with 3 levels for each parameter on the input-output matrices of LOLIB, it is determined that the best values are: *TabuTenure* $= 2\sqrt{n}$, *MaxIter* $= n$ and *MaxDiv* $= 0.5n$. In the second computational experiment the authors compared:

-   the short term memory tabu search method, STS (without the long-term diversification,
-   the complete tabu search method, TS,
-   the method of Chanas and Kobylanski [31],
-   the method of Becker [12], and
-   the improvement method of the GRASP algorithm, LS, described in Sect. 3.2.

We refer to Chanas and Kobylanski's method as CK, and as CK-10 for the application of the method from 10 randomly generated initial solutions. In a similar way, LS-10 refers to the application of the improvement local search method from 10 different starting solutions. The experimentation in [106] is limited to three sets of instances: the input-output instances IO of LOLIB, the instances SGB from the Stanford GraphBase [99] and randomly generated instances, RandomAI. A uniform distribution with parameters (0, 25000) was used to generate the random instances of sizes 75, 150 and 200 (25 instances per size). Table 3.3 shows, for each method, the average percentaged deviation from optimality and/or best known solutions, the number of optimum solutions, and the average CPU time (seconds on a 166 MHz Pentium). (Since the optimum solutions were not known for the random instances, we list in this case the number of best solutions found.)

Table 3.3 Comparison of methods

	LS	LS-10	Becker	CK	CK-10	STS	TS
Input-output tables							
Deviation	0.15	0.02	8.95	0.15	0.02	0.04	0.00
Normal deviation	0.18	0.03	10.38	0.18	0.03	0.05	0.00
No. of opt. solutions	11	22	0	11	27	30	47
CPU time	0.01	0.08	0.02	0.10	1.06	0.33	0.93
SGB instances							
Deviation	0.18	0.01	2.04	0.08	0.01	0.02	0.00
No. of opt. solutions	3	20	0	4	22	35	70
CPU time	0.06	0.55	0.20	1.45	16.33	1.16	4.09
Random instances							
Deviation	0.47	0.23	3.08	0.53	0.28	0.06	0.00
No. of best solutions	0	2	0	0	0	40	73
CPU time	0.12	1.21	0.80	10.67	108.44	10.79	20.19

Table 3.3 shows that Becker's procedure is clearly inferior in terms of solution quality although, given its simplicity, its performance is quite acceptable. The performance of the LS and CK methods is very similar across the three problem sets. TS outperforms all other methods in terms of solution quality, specially in the Random set, where it provides 73 of the 75 best known solutions. However, in terms of computational effort, TS consumes more running time than the other methods, with the exception of CK-10. The short term memory implementation, STS, performs very well considering its short running time. For the input-output matrices, we report in Table 3.3 both the deviation reported in [106] (see row "Deviation") and the deviation computed for the instances transformed to normal form (see row "Normal deviation"). As expected the latter is slightly larger (we include both deviations as a baseline for future comparisons).

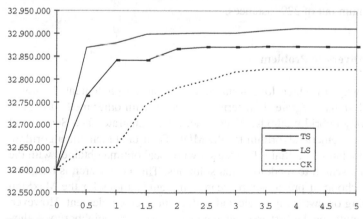

Fig. 3.5 Performance graph

As a final result of the study in [106] the performance graph shown in Figure 3.5 is given in which the best three methods identified in the previous experiment were run in a way that the best solution found was reported every 0.5 seconds. These data points were used to generate the graph. The difference in quality between CK and the LS procedure may be due to the number of initial solutions used for each method. While the LS procedure can be applied 150 times during 5 seconds, the CK method can be applied only 10 times during the same amount of time. The superior performance of TS is made evident by Figure 3.5.

This study reveals that, although ignored in some tabu search implementations, the long term diversification by itself is an important component of the methodology, since in all cases it enhanced the performance of the basic procedure (as can be seen in Table 3.3 comparing the results of STS and TS).

We now report on our own experiment with tabu search for the LOP. Specifically, we apply the tabu search algorithm for 10 seconds to the 229 instances in the OPT-I set (with optimum known). Table 3.4 shows the number of instances in each set, the average percentage relative deviation, *Dev*, between the best solution value found with this method and the optimal value, and the number of instances, *#Opt*, for which an optimum solution is found.

Table 3.4 Tabu search on OPT-I instances

	IO	SGB	RandAII	Rand B	MB	Special	Total
#Instances	50	25	25	70	30	29	229
Dev(%)	0.00	0.00	0.00	0.00	0.00	0.02	0.00
#Opt	50	25	25	70	30	15	215

The results in Table 3.4 clearly show that the tabu search method is able to obtain the optimal solution for most of the instances in the OPT-I set. Specifically, it obtains 215 optimum out of 229 instances.

The Maximum Diversity Problem

There are many ways in which long term strategies can be implemented. Nowadays, advanced designs integrate short term tabu search with other methodologies, creating in this way hybrid methods. Wang et al. [162] integrate Tabu Search and Scatter Search in a memetic algorithm for the MDP. Their tabu memetic algorithm, called TS-MA populates an initial reference set with local optima obtained with the application of tabu search to random initial solutions. This tabu search is based on the same neighborhood of previous tabu search implementations for the MaxSum problem, consisting on swapping a selected with an unselected element. However, to reduce the computational effort associated with exploring the neighborhood, they apply a successive filter candidate list strategy, and subdivide the move into its two natural components: first remove an element, and then add another element. The authors explain that one of the key elements in their memetic algorithm is the solution combination operator based on solution properties by reference to the analysis of strongly determined and consistent variables. The method performs iterations combining the solutions in the reference set as long as the resulting solutions qualify to enter to this set.

Palubeckis [137] implemented a simple and efficient long term strategy for his Iterated Tabu Search (ITS). In particular, the algorithm repeatedly invokes two procedures, a short term memory tabu search, and a perturbation method for diversification purposes. The short term phase includes the so-called tabu list to forbid flipping the values of recently changed binary variables (i.e., changing them from 0 to 1, or from 1 to 0). This phase is coupled with a standard local search to reach the corresponding local optimum from the current solution. The perturbation phase is then applied to generate a new initial solution far from the recently explored region. Specifically, p components of the binary vector are randomly selected, where $p \leq 2 \min(m, n - m)$ from a short candidate list of components with relatively large distance values. The selected components are flipped to obtain a feasible solution, and the method applies the short term tabu search again from it.

3.5 Simulated Annealing

The *simulated annealing* methodology [98] is based on a presumed analogy between the physical process known as annealing of solids and the algorithmic process of solving an optimization problem.

Annealing refers to a physical process in which a solid, placed in a heat bath, is first heated up by increasing the temperature of the bath, and then cooling down by slowly lowering this temperature. In this way, the particles of the solid arrange themselves in the low energy ground state (called crystal). If the cooling is done very quickly, irregularities are locked into the particles' structure, thus obtaining an amorphous solid (with a higher trapped energy level than in a perfectly structured crystal). Therefore, to obtain a perfect crystal, the temperature must be lowered gradually, following what Kirpatrick et al. [98] called *careful annealing*, where temperature descends slowly through a series of levels. At each temperature value t, the solid is then allowed to reach *thermal equilibrium*, which can be characterized by a probability given by the Boltzmann distribution [48].

From the optimization viewpoint, SA proceeds in the same way as ordinary local search but incorporates some randomization in the move selection process. Specifically, it avoids getting trapped in a local optimum by means of nonimproving moves. These moves are accepted according to certain probabilities taken from the analogy with the annealing process. To simulate the change to thermal equilibrium of a solid for a given value of t, Metropolis et al. [124] proposed a *Monte Carlo method*. This simulation, introduced in the late 1940's to model the behavior of gases in a heat bath, is implemented in SA to determine whether a change in the physical system will be accepted or not.

Simulated annealing claims that an optimization problem or more precisely, the set of its feasible solutions, can be interpreted as the states of a physical system and that the process of careful annealing can be re-interpreted for the optimization problem to find an optimum solution. An optimum solution thus corresponds to a ground state. This analogy is shown in Table 3.5.

Table 3.5 Simulated annealing analogy

Physical system	Optimization problem
State	Feasible solution
Energy	Objective function
Ground state	Optimum solution
Careful annealing	Simulated annealing

At a given state of the solid (solution), the Monte Carlo method applies a random small perturbation and computes the difference in energy ΔE between the current state and the perturbed state. If this difference is negative ($\Delta E < 0$), the change is accepted and the perturbed solution becomes the current one because it has lower energy. However, if this difference is positive ($\Delta E > 0$), the probability of the perturbed state being accepted is computed according to the Boltzmann distribution. In mathematical terms

$$P(\text{new solution is accepted}) = e^{-\Delta E / t k_B},$$

where k_B is the *Boltzmann constant* [48]. The Monte Carlo method (also called the *Metropolis algorithm*) is applied to generate moves (states) and accept all improving moves, but in addition also worsening moves with a certain probability. It should be noted that, when run long enough, Monte Carlo simulation indeed generates states according to the Boltzmann distribution.

As is tabu search (TS) in its basic form, simulated annealing is also based on local search. When comparing SA and TS with a simple descent method to maximize $f(x)$, we know that such a method only permits improving moves to neighbor solutions. In physical terms, this would be *rapid quenching* instead of careful annealing. Simulated annealing and tabu search, on the other hand, permit moves that deteriorate the current objective function value.

In the previous section we described how the moves are chosen in TS from a modified neighborhood $\mathcal{N}^*(x)$ according to the tabu status. Non-improving moves are accepted in SA according to the Boltzman distribution. Since the LOP is a maximization problem, a move from x to y is an improving one if $\Delta = f(y) - f(x) > 0$.

The basic outline of a simulated annealing method is the following. The algorithm starts at some temperature t which is decreased in the course of the algorithm. For each temperature, a number of moves according to the Metropolis algorithm is executed to simulate getting to the thermal equilibrium. Various schemes for guiding this process are given in the literature. For certain schemes one can prove that this algorithm indeed yields an optimum solution with probability 1, but these schemes lead to much too slow a convergence to be useful for practical applications. But at least, this result indicates the potential of the approach.

Simulated Annealing

(1) Generate an initial solution x.

(2) Choose an initial temperature t, a cooling factor c, $0 < c < 1$, and a repetition factor r.

(3) While not *frozen*:

    (3.1) Repeat the following steps r times:

        (3.1.1) Choose $y \in \mathcal{N}(x)$ at random.

        (3.1.2) Compute $\Delta = f(y) - f(x)$.

        (3.1.3) If $\Delta \geq 0$, set $x = y$.

        (3.1.4) If $\Delta < 0$, compute a random number p, $0 \leq p \leq 1$. If $p \leq \frac{e^{\Delta}}{t}$, set $x = y$.

(4) Update temperature $t = ct$.

Johnson et al. [90] give some guidelines to set the values of the parameters in the SA algorithm shown above: the initial temperature t, the cooling factor c, the repetition factor r and how to establish the stopping criterion (when it is *frozen*). The authors declare themselves to be skeptical about the relevance of the details of the analogy to the actual performance of SA algorithms in practice (and we share their point of view). In line with this, they examine and test the values of this key search parameters from the performance viewpoint, regardless of their meaning in the physical process. They actually translate the computation of these parameters into others that are more closely related to the optimization process:

- A new parameter, *initprob*, is used to determine an initial temperature t. Based on an abbreviated trial annealing run, a temperature is found at which the fraction of accepted moves is approximately *initprob*, and this is used as a starting temperature. They recommend setting *initprob* to 0.4 in the problem tested.
- The repetition factor r is set as $r = 16s$ where s is the expected neighborhood size.
- The *minpercent* parameter is introduced to determine whether the annealing run is *frozen* or not. A counter is kept which is incremented each time a temperature is completed for which the percentage of accepted moves is at most *minpercent*, and is reset to 0 each time a solution is found to be better than the incumbent one. If the counter reaches 5 the process is declared *frozen* and the SA algorithm stops.

As far as we know there is no previous implementation of the SA methodology to the LOP. However, there are several approximations with similar methods or variants of the LOP. In [15] a SA method is proposed for the *chromosome reconstruction problem*. This problem arises when creating maps in genetics to reconstruct the

DNA sequence. Although the authors called the mathematical formulation of this problem *optimal linear ordering*, it should be noted that it involves computing an ordering (clone ordering) that minimizes the sum of differences between successive clones. However, in this computation they do not include the differences between noncontiguous elements (clones) and only consider consecutive elements. Thus, it is not the same objective function as in the LOP. On the other hand, as will be shown in a subsequent chapter, SA has been implemented to solve an auxiliary problem that appears when studying the convex hull of feasible solutions (the polytope) of the LOP. Specifically, we will see how this method is applied to separate small facets defining inequalities for the LOP.

Charon and Hudry [33] describe the application of the *noising method* to some combinatorial optimization problems, including the LOP. This method proceeds in the same way as local search, performing a move at each iteration, from the current solution x to a neighbor solution $y \in \mathcal{N}(x)$. However, when they compute the *move value*, $\Delta = f(y) - f(x)$, they add a perturbation or *noise* and, instead, consider $\Delta_{noise} = f(y) - f(x) + \sigma$ where σ is a random value drawn from the interval $[-r, r]$ according to a pre-established probability distribution. As the search progresses the *noise rate* r decreases and the method finishes when no improvement move is available. The authors point out that different methodologies, including SA, can be considered an instance of the noising method. In other words, if we choose the parameters properly, especially the probability distribution, the noising method generalizes these methodologies. In their computational experiments, the authors test their adaptation of the noising method to the LOP with different parameters, including noising rates, descent variants and restarting mechanisms, and conclude that the method provides high quality solutions.

As in the previous sections, we report the results obtained when applying this method for 10 seconds to the 229 instances in the OPT-I set. Specifically, we implemented the SA method described in the basic outline above. Table 3.6 shows the number of instances in each set, the average percentage deviation, *Dev*, and the number of instances, *#Opt*, for which SA obtains an optimum solution.

Table 3.6 Simulated Annealing on OPT-I instances

	IO	SGB	RandAII	Rand B	MB	Special	**Total**
#Instances	50	25	25	70	30	29	229
Dev(%)	0.03	0.03	0.08	0.25	1.33	0.39	0.35
#Opt	16	0	0	10	0	6	32

The results in Table 3.6 clearly show that the simulated annealing obtains good solutions on the OPT-I set, since the average percentage deviation overall is 0.35%. However, these results are inferior in quality to those obtained with the tabu search method (as shown in Table 3.4, TS presents an average percent deviation of 0.00%). At the end of this chapter we will show an extensive comparison among all the methods described.

3.6 Variable Neighborhood Search

Variable neighborhood search (VNS) is rapidly becoming a method of choice for designing solution procedures for hard combinatorial optimization problems [80]. VNS is based on a simple and effective idea: a systematic change of the neighborhood within a local search algorithm. In this section we follow the description given in [62] to adapt the VNS methodology to the LOP.

As stated in [80], VNS is based on the following three principles, where principle (2) is true for all optimization problems, but principles (1) and (3) may or may not hold depending on the problem at hand.

(1) A local minimum with respect to one neighborhood is not necessarily so with another.
(2) A global minimum is a local minimum with respect to all possible neighborhood structures.
(3) For many problems local minima with respect to one or several neighborhoods are relatively close to each other.

To apply the VNS methodology we first need to define different neighborhoods for our problem. In the LOP, let again $move(p_j, i)$ denote the local modification where p_j is deleted from its current position j in permutation p and inserted at position i (i.e., between the objects currently in positions $i-1$ and i). In [62] García et al. proposed the following neighborhoods to adapt the VNS methodology to the LOP:

$$\mathcal{N}_1(p) = \{p' \mid p' \text{ is obtained by } move(p_j, i), \text{ for some } j = 1, \dots, n,$$
$$\text{and } i \in \{j-1, j+1\}\},$$
$$\mathcal{N}_2(p) = \{p' \mid p' \text{ is obtained by } move(p_j, i), \text{ for some } j = 1, \dots, n,$$
$$\text{and } i = 1, \dots, n, i \neq j\}.$$

The neighborhood $\mathcal{N}_1(p)$ consists of permutations p' that are reached by switching the positions of contiguous objects in p. $\mathcal{N}_2(p)$ consists of all permutations p' resulting from executing general insertion moves in p. Based on $\mathcal{N}_2$, composed neighborhoods $\mathcal{N}_k$, $k = 3, \dots, n$, can be defined. The neigborhood $\mathcal{N}_k(p)$ is the set of solutions that are obtained when we apply the general insertion move $k-1$ times from p. Obviously, $\mathcal{N}_k$ can be defined in terms of $\mathcal{N}_2$. E.g., for $k = 3$, a solution r is in $\mathcal{N}_3(p)$ if, for some $q \in \mathcal{N}_2(p)$, we have $r \in \mathcal{N}_2(q)$.

In the next subsection we will describe five different implementations of variable neighborhood search (and its variants) for the LOP, and then report some computational experiments to compare them. Then, we will describe a competitive VNS implementation for the MDP.

3.6.1 VNS implementations for the LOP

The **Variable Neighborhood Descent (VND)** method is obtained when the change between neighbors is performed in a deterministic way. The implementation for the LOP performs a local search for the best solution in $\mathcal{N}_1$ and only resorts to performing one move in $\mathcal{N}_2$ when the search is trapped in a local optimum found in $\mathcal{N}_1$. If the local search in $\mathcal{N}_2$ is able to improve the best solution found so far, the search continues in $\mathcal{N}_1$, otherwise it terminates.

The **Restricted Variable Neighborhood Search** RVNS implementation for the LOP is also limited to $\mathcal{N}_1$ and $\mathcal{N}_2$. It repeatedly performs three steps combining stochastic and deterministic strategies (assume $k \in \{1,2\}$):

(1)   A solution p' is randomly generated in $\mathcal{N}_k(p)$. (*Shaking*)
(2)   Apply a local search from p' to obtain a local optimum p''. (*Improving*)
(3)   If p'' is better than p, then p is replaced by p'' and k is set to 1; otherwise, k is switched (from 1 to 2 or from 2 to 1). (*Updating*)

As in VND, k is initially set to 1 and the method resorts to $\mathcal{N}_2$ when $\mathcal{N}_1$ (now in combination with local search) fails to improve the current solution. However, if $\mathcal{N}_2$ also fails to improve on the incumbent solution, instead of stopping the search, RVNS sets $k = 1$ and randomly selects another trial solution in $\mathcal{N}_1$, repeating the three steps again. The sequence is repeated until a *MaxIter* number of consecutive iterations is performed with no further improvement. In step (2) we apply the local search method based on $move(p_j, i)$ ($\mathcal{N}_2(p)$) described in the previous chapter.

The **Basic Variable Neighborhood Search** method (BVNS) follows the same scheme as RVNS based on the three steps shaking, improving and updating. However, in this version the method uses $k_{\max}$ neighborhoods.

Basic variable neighborhood search

(1)   Generate an initial solution p.
(2)   Define neighborhoods $\mathcal{N}_k$ with $k=1,\dots,k_{\max}$.
(3)   Set *counter* $= 0$ and $k = 1$.
(4)   While *counter* $<$ *MaxIter*:

    (4.1)   Randomly generate a solution p' in $\mathcal{N}_k(p)$.
    (4.2)   Apply a local search from p' to obtain a local optimum p''.
    (4.3)   If p'' is better than p, then set $p = p''$, *counter* $= 0$ and $k = 1$.
    (4.4)   Else set $k = k + 1$ (if $k = k_{\max}$, then set $k = 1$) and *counter* $=$
    counter $+ 1$.

Initially, k is set to 1 and in the shaking step a solution $p' \in \mathcal{N}_k(p)$ is randomly generated. Then, a local search method is applied from p' to obtain a local optimum p''. In the update step, if p'' is better than p, p is replaced by p'' and k is set to 1; otherwise, k is incremented. The method repeats these three phases until a *MaxIter* number of consecutive iterations is performed with no further improvement. As in the previous version, we consider the $\mathcal{N}_2$ descent procedure as the local search phase.

The **Frequency Variable Neighborhood Search** for the LOP combines VNS and tabu search. As we have seen in the previous variants, VNS is mainly based on random sampling of selected neighborhoods in combination with local search. On the other hand, tabu search is based on the notion of recording information (attributes) to perform a guided and deterministic (i.e., not random) search of the solution space. We therefore can say that to some extent they represent opposite approaches. However, in the frequency variable neighborhood search, FVNS, both approaches are integrated into a single design.

As in BVNS, the FVNS method repeats shaking, improving and updating steps until a *MaxIter* number of consecutive iterations is performed with no further improvement. However, in FVNS, the shaking step is guided by frequency information recorded in previous iterations. This is made to diversify the search in a controlled way (instead of a completely random and uncontrolled way). As previously shown (see Sect. 3.3), diversification is the notion of expanding the search to unexplored regions in the solution space (and it does not necessarily means randomization). Diversification strategies are generally based on either encouraging the incorporation of new elements or discouraging elements visited frequently.

In Chap. 2 we compared several different constructive methods for the LOP. Among them, in particular, FQ is based on a frequency counter that records the number of times an element appears in a specific position. This counter is used to penalize the "attractiveness" of an element with respect to a given position. In a similar way, we now use a frequency counter $freq(i)$ to record the number of times object i has been moved. Therefore, each time object i is moved from one position to another in the shaking or the local search phase, we increment $freq(i)$ by one unit. We use this frequency counter to generate a new solution in the shaking step. Since we want to diversify, we select objects j with a small frequency value $freq(j)$. Specifically, in the shaking step, we randomly select $k_{\max}$ objects in the incumbent solution p to be moved. The probabilistic selection rule is inversely proportional to the frequency count. The selected objects are moved to the best available position.

As mentioned in the introduction of this chapter, researchers in the meta-heuristic field increasingly propose combinations and hybridizations among the different methodologies, which makes the boundary between them less well defined, in many cases taking some elements from others. Actually we described in the previous subsection FVNS in which a simple memory structure based on frequencies (a tabu search element indeed) plays an important role in the shaking step of VNS. We describe now two straightforward *Hybridizations of VNS* in which we simply replace the local search of the improving step with a meta-heuristic method, thus obtaining the following three steps:

(1) A solution p' is randomly generated in $\mathcal{N}_k(p)$. (*Shaking*)
(2) Apply a meta-heuristic from p'. Let p'' be the output solution. (*Searching*)

(3) If p'' is better than p, then p is replaced by p'' and k is set to 1, otherwise, k is
     incremented by 1 (if $k = k_{max}$, k is set to 1). (*Updating*)

We considered the two previous VNS variants, BVNS and FVNS, and replaced
the local search with the short term memory tabu search algorithm described in the
previous section (in which no longer term structures are present) according to the
scheme above. We called the resulting methods BVNS-TS and FVNS-TS, respectively. In their computational experiments, García et al. [62] tested these procedures
on the input-output instances from LOLIB, the Stanford GraphBase instances, and
the random instances A of type I and type II.

In a preliminary experiment of [62] simple local search methods are compared
to find appropriate values of the key search parameters. Specifically, they compare
VND with the local search based on the $\mathcal{N}_2$ neighborhood on both real and random instances. The experiment confirms what is well known for the LOP: random
instances are more difficult than the real input-output instances. It also shows that
there are small variations in the results of these procedures. However, if we run them
from different initial solutions, it becomes clear that VND saves time since it only
resorts to $\mathcal{N}_2$ when the search is trapped in $\mathcal{N}_1$.

In the first experiment of [62] VNS variants that do not incorporate long term
strategies are compared with two previous methods on the input-putput instances
and on the random instances of type I. Specifically, they compare RVNS, BVNS,
FVNS with the short term tabu search, ST-TS [106], described in the previous section, and the method by Chanas and Kobilansky, CK-10 [31], run from 10 randomly
generated initial solutions. Table 3.7 reports the number of best solutions found (optimum solutions for the LOLIB) and the percentaged deviation of each method on
each set of instances.

Table 3.7 shows that the best solution quality is obtained by the variants of the
VNS methodology, which are able to match a larger number of optimum and best
known solutions than the short term TS and CK methods. In particular, for the input-
output instances, RVNS finds 39 optimum solutions, BVNS 35, FVNS 34, ST-TS 30
and CK-10 finds 27. On the other hand, on the random instances, RVNS determines
only 10 best known solutions out of 25, BVNS 19, FVNS 17, ST-TS 14 and CK-10
finds 4. All methods are extremely fast considering that their running times are
below 0.02 seconds. The performance of CK is clearly inferior with a lower number
of optimum solutions than those achieved by the other approaches. However, as
mentioned, it is a simple heuristic and its results are quite acceptable considering its
simplicity.

Table 3.7 Comparison of basic methods

Instances	RVNS	BVNS	FVNS	ST-TS	CK-10
Input-output					
No. of opt. solutions	39	35	34	30	27
Percentaged deviation	0.02	0.03	0.05	0.04	0.02
Random A type I					
No. of best solutions	10	19	17	14	4
Percentaged deviation	0.15	0.03	0.04	0.05	0.12

The final experiments in [62] compare basic and hybrid VNS with the best heuristics for the LOP, in particular BVNS, BVN-TS, FVN-TS and the previous approaches TS (the complete tabu search approach described in the previous section) and SS (the scatter search approach described in the next section).

Table 3.8 shows, for each procedure, the average deviation from optimality in percent, the number of optimum solutions, and the average CPU time in seconds for each set of instances. Since optimum solutions are not known for the large random instances, the deviation for these problems is given with respect to the best solution found during each experiment. Also the number of best solutions found is reported instead of the number of optimum solutions. We have set the stopping parameter *MaxIter* in the VNS versions to 100 to approximate the running time consumed by the LT-TS method.

For the input-output instances the long term tabu search algorithm TS is able to obtain 47 optimum solutions in 0.024 seconds while the VNS variants obtain a number of optimum solutions that ranges between 39 and 46 in less computation time than LT-TS. The performance of the SS method in this experiment is clearly inferior in terms of quality considering its running time.

For large and more difficult instances, as in the previous experiments, SS obtains very good solutions but it needs longer running times than the other methods. FVN-TS and LT-TS are clearly the best methods in terms of solution quality achieved within small running times. Both obtain the same number of optimum solutions for the SGB instances, although LT-TS presents a smaller percentaged deviation on average and a larger computational time than FVN-TS. On the other hand, FVN-TS obtains 6 best solutions and 0.16 average percent deviation in the random type I instances, while LT-TS obtains 10 best solutions and 0.06 average percent deviation. Results in random type II instances are different since FVN-TS is able to obtain 29 best solutions in 0.22 seconds of running time, which compares favorably with all the other methods considered.

It is interesting to see that, although the frequency VNS version (FVNS) does not improve the memory less variant (VNS) as shown in previous tables, when we coupled the VNS methods with tabu search it seems that the use of frequency based memory improves the basic VNS in solving the LOP.

Table 3.8 Comparison of best methods

	BVNS	BVN-TS	FVN-TS	TS	SS
Input-output instances					
Deviation	0.0208	0.0370	0.0082	0.0007	0.0133
No. of opt. solutions	40	41	46	47	42
CPU time	0.015	0.013	0.018	0.024	0.04
SGB instances					
Deviation	0.0251	0.0087	0.0104	0.0018	0.0023
No. of opt. solutions	7	11	14	14	15
CPU time	0.067	0.039	0.052	0.090	0.153
Random A type I					
Deviation	0.1870	0.1615	0.1600	0.0615	0.0130
No. of best solutions	4	5	6	10	48
CPU time	1.020	0.289	0.305	0.417	0.709
Random A type II					
Deviation	0.0053	0.0029	0.0019	0.0014	0.0017
No. of best solutions	16	20	29	28	18
CPU time	0.607	0.336	0.220	0.269	0.457

Figure 3.6 shows in a box-and-whisker plot the value of the best solution obtained with LT-TS and FVN-TS on the largest random instances (types I and II with n=200).

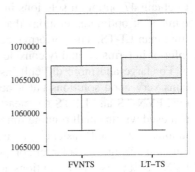

Fig. 3.6 Best value box plot

Note that the higher the value, the better the method (since we are maximizing the objective function). This figure clearly shows that both methods present similar results (in terms of the best solution found) although LT-TS presents a marginal improvement over FVN-TS in these instances since it is able to obtain, in some cases, better solutions (the higher *whisker* is larger in LT-TS than in FVN-TS).

As in previous sections, we report the results obtained when applying this method for 10 seconds to the 229 instances in the OPT-I set. Specifically, we consider the basic VNS method described above. Table 3.9 shows the number of instances in each set, the average percentaged deviation *Dev*, and the number of instances *#Opt*, for which SA obtains an optimum solution.

Table 3.9 BVNS on OPT-I instances

	IO	SGB	RandAII	Rand B	MB	Special	**Total**
#Instances	50	25	25	70	30	29	229
Dev(%)	0.00	0.00	0.00	0.02	0.00	0.11	0.02
#Opt	50	25	19	64	30	20	208

Results in Table 3.9 clearly show that the VNS methodology obtains high quality solutions on the OPT-I set. The average percentage deviation overall is 0.02% and the number of optimum solutions is 208 out of 229 instances.

A very recent study on the LOP from 2020 includes a theoretical analysis of the structure of its solutions, and its application in a VNS method. In particular, Santucci and Ceberio [152] review previous LOP solving methods and conclude that the insert move has been extensively applied in local search based heuristics, and it has exceptional properties that can be further exploited. The authors consider two key features that help them to create a very efficient VNS:

- Given an ordering or permutation $O = \langle O_1, O_2, \ldots, O_n \rangle$ that describes a solution for a LOP instance, where the element O_i is in position i, the ordering of the previous and posterior elements with respect to i, does not affect its contribution to the objective function.
- The objective value of a solution is only given by the pairwise precedences among the permuted elements in the solution.

As pointed out by the authors, representing a linear ordering as a set of precedences requires more memory than a linear permutation encoding. However, this representation allows a better control of its value. Specifically, they propose to express the objective function as a single summation of elementary coefficient matrix elements. This permits to design constructive heuristics that smoothly build-up a solution, avoiding drastic changes typical for previous methods.

Santucci and Ceberio implemented a restricted neighborhood to speed up the move selection process. It is based on the analysis [29] that for each element in a permutation, there is a set of positions, regardless the ordering of the other elements, in which the element cannot be in a LOP local optimum. We elaborate more on this property in Section 3.8 on Genetic and Memetic algorithms, where it is applied on that type of methods.

The VNS method [152] implements two neighborhoods, insertions and exchanges, in a restricted exploration due to the property mentioned above. It is coupled with a construction/destruction heuristic working with the precedences set representation, that generates starting points to launch the VNS. This heuristic implicitly implements a long term memory structure, since it removes from the solution (output of the VNS) precedences that have been met more often in the local optima with diversification purposes. Then, the constructive part rebuilds the solution, adding precedences, to feed the VNS in a new iteration. The method is tested on very large instances (with up to 8000 elements) and compared with the ILS [154].

The comparison favors the proposed method, and also identifies new best-known solutions.

3.6.2 VNS implementations for the MDP

Silva et al. [153] proposed in 2004 a simple VNS algorithm, SOMA, for the MDP. The method, based on two neighborhoods, first applies the classic local search for the MDP based on exchanges until no further improvement is possible. Then, a local search based on an expanded neighborhood is executed. The new neighborhood is defined as the set of all solutions obtained by replacing two elements in the solution (selected elements) with another two that are not present in the current solution (unselected elements). Considering that a solution of the MDP has a fixed number of selected elements, m, a move can be defined as the exchange of an arbitrary number of elements $k \leq m$ in the solution with the same number of elements not in the solution. This leads to a standard neighborhood definition in binary problems. Additionally, note that VNS typically implements nested neighborhoods, which in this case are naturally defined in this way.

Brimberg et al. [20] proposed several VNS procedures originally devoted to the heaviest k-subgraph problem, which is a generalization of the MDP in which the distances are replaced with weights (that can take arbitrary values, including 0). Therefore, any algorithm designed to the k-subgraph is able to solve the MDP. The authors presented different VNS versions, including *Skewed VNS*, basic VNS and a combination of a constructive heuristic followed by VNS. The best overall method according to their experimentation is the basic VNS, B-VNS, which consists of three main elements. The first one, called Data Structure, allows the algorithm to efficiently update the value of the objective function; the second one, Shaking, generates solutions in the neighborhood of the current solution by performing random vertex swaps; and third one is a local search procedure based on exchanges.

In 2011, Aringhieri and Cordone [8] presented four implementations of the VNS methodology to solve the MDP. They are called Basic VNS, Guided VNS, Accelerated VNS and Random VNS. In all of them, the initial solution is constructed with a greedy method similar to the constructive methods described in Chapter 2. Given a solution, each iteration in any of the four methods consists of generating a new solution by replacing k elements in the current solution with k elements out of it (Shaking procedure). The new solution is improved with a basic Tabu Search. Only if the new solution is better or sufficiently distant than the previous one, it becomes the current solution; otherwise it is discarded, and a new solution is generated with the same replacement mechanism in which k is augmented by one unit. The methods finish when a maximum number of iteration is reached or k reaches a pre-established value *kmax*.

In the *Basic VNS*, the Shaking procedure randomly unselects k elements in the current solution and randomly selects another k out of it to replace them. In the *Guided VNS*, the shaking procedure is deterministic and the elements are selected/unselected according to a frequency value, which records the number of times that each element has been included in previous solutions. The *Accelerated VNS* modifies the basic variant by setting $kmax = min(m, n-m)$. This strategy makes re-starts much less frequent because $kmax$ is considerably larger than the values used in the Basic and Guided variants. Finally, the *Random VNS* sets $kmin = kmax = min(m, n-m)$, which means that the procedure always re-starts the search in the largest neighborhood, instead of gradually enlarging the neighborhood used. This approach corresponds to a nearly random restart, which only forbids the elements belonging to the current best-known solution. Experimental results show that the random VNS outperforms the other three variants.

3.7 Scatter Search

Scatter search (SS) is an evolutionary or population based method in the sense that it operates on a set of solutions, combining them to obtain new and hopefully better solutions [105]. Nowadays it is a well established method within the meta-heuristic community and has been successfully applied to a wide range of optimization problems. However, general awareness of the method still lags behind that of other population based methods such as genetic algorithms or well established meta-heuristics like tabu search.

There are three elements that we need to define in any evolutionary method: a way to generate solutions, a way to combine solutions and a way to maintain a set (population) of solutions. When we design these elements for the problem we are faced with, we follow the guidelines given by the meta-heuristic. In this section we will see how to define these elements for the LOP and the MDP, and how they interact according to the SS methodology.

Scatter search was first introduced in [70] as a heuristic for integer programming. In the original proposal, solutions are purposely (i.e., non-randomly) generated to take characteristics of various parts of the solution space into account. Scatter search orients its explorations systematically relative to a set of reference points that typically consist of good solutions obtained by prior problem solving efforts, where the criteria for "good" are not restricted to objective function values, and may apply to subcollections of solutions rather than to a single solution, as in the case of solutions that differ from each other according to certain specifications.

The scatter search methodology is very flexible, since each of its elements can be implemented in a variety of ways and degrees of sophistication. In this section we give a basic design to implement scatter search based on the following five methods:

(1) A *diversification generation method* to generate a collection of diverse trial solutions, using an arbitrary trial solution (or seed solution) as an input.
(2) An *improvement method* to transform a trial solution into one or more enhanced trial solutions. (Neither the input nor the output solutions are required to be feasible, though the output solutions will usually be expected to be feasi-

ble. If no improvement of the input trial solution results, the enhanced solution is considered to be the same as the input solution.)

(3)   A *reference set update method* to build and maintain a reference set consisting of the b best solutions found (where the value of b is typically small, e.g., not greater than 20), organized to provide efficient access by other parts of the method. Solutions gain membership to the reference set according to their quality or their diversity.

(4)   A *subset generation method* to operate on the reference set and to produce a subset of its solutions as a basis for creating combined solutions.

(5)   A *solution combination method* to transform a given subset of solutions produced by the subset generation method into one or more combined solution vectors.

In Figure 3.7, diversification generation and improvement methods are initially applied, adding improved solutions to P, until the cardinality of P reaches *PSize* solutions that are different from each other. The darker circles represent improved solutions resulting from the application of the improvement method. The main search loop appears to the left of the box containing the reference solutions (labeled *RefSet*). The subset generation method takes reference solutions as input to produce solution subsets to be combined. Solution subsets contain two or more solutions. The new trial solutions resulting from the application of the combination method are subjected to the improvement method and handed to the reference set update method. This method applies rules regarding the admission to the reference set of solutions coming from P or from the application of the combination and improvement methods. Of the five methods in scatter search, only four are strictly required. The improvement method is usually needed if high quality outcomes are desired, but a scatter search procedure can be implemented without it. The advanced features of scatter search are related to the way the five methods are implemented. That is, the sophistication comes from the implementation of the SS methods instead of the decision to include or exclude some elements (like in the case of tabu search).

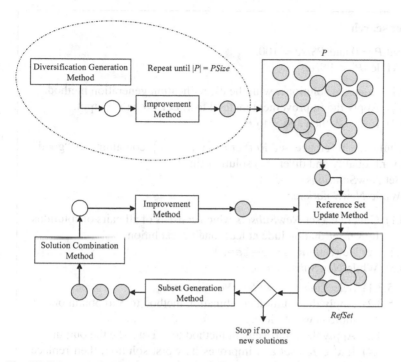

Fig. 3.7 Schematic flow diagram of scatter search

Similarities and contrasts between SS and the original proposals for genetic algorithms (GA) are observed in [105]. Both are instances of what are sometimes called "population based" or "evolutionary" approaches. Both incorporate the idea that a key aspect of producing new elements is to generate some form of combination of existing elements. However, GA approaches are predicated on the idea of choosing parents randomly to produce offsprings, and further on introducing randomization to determine which components of the parents should be combined. In contrast, the SS approach does not emphasize randomization, particularly in the sense of being indifferent to choices among alternatives. Instead, the approach is designed to incorporate strategic responses, both deterministic and probabilistic, that take evaluations and history into account. SS focuses on generating relevant outcomes without losing the ability to produce diverse solutions, due to the way the generation process is implemented.

The core of the SS method is the *reference set* in which good solutions are stored (where good refers not only to quality but also to diversity). The scatter search algorithm has three main stages according to the status of the reference set: reference set creation, reference set update, and reference set rebuild. The following pseudo-code shows how to implement a basic Scatter Search algorithm.

Scatter search

(1) Set $P = \emptyset$ and $PSize = 100$.
(2) While $|P| < PSize$:

    (2.1) Generate solution x with the diversification generation method.
    (2.2) Apply the improvement method to x. Let x' be the output.
    (2.3) If $x' \notin P$, then add x' to P.

(3) Build the reference set $RefSet = \{x_1, \ldots, x_b\}$ containing b "good" (w.r.t. quality and diversity) solutions in P.
(4) Set $NewSolutions = $ True.
(5) While $NewSolutions$:

    (5.1) Generate a set $NewSubsets$ which consists of all pairs of solutions in $RefSet$ that include at least one new solution.
    (5.2) Set $NewSolutions = $ False.
    (5.3) While $NewSubsets \neq \emptyset$:

        (5.3.1) Select the next set $S \in NewSubsets$.
        (5.3.2) Apply the solution combination method to S to obtain one or more new solutions x.
        (5.3.3) Apply the improvement method to x. Let x' be the output.
        (5.3.4) If $x' \notin RefSet$ and improves its worst solution, then replace this with x' and set $NewSolutions = $ True.
        (5.3.5) Delete S from $NewSubsets$.

(6) Output the best solution in $RefSet$.

In the following subsections we will describe how the three main SS stages can be implemented to build a scatter search algorithm for a given problem, and how they are adapted in [24] to the LOP and in [61] to the MDP.

3.7.1 Reference Set Creation

The diversification generation method is used to build a large set P of diverse solutions. The size $PSize$ of P is typically at least 10 times the size of $RefSet$. The initial reference set is built taking solutions from P according to the reference set update method.

The reference set $RefSet$ is a collection of both high quality solutions and diverse solutions, which are used to generate new solutions by way of applying the combination method. In this basic design we can use a simple mechanism to construct an initial reference set based on a distance function between solutions, and then update it during the search. The size of the reference set is denoted by $b = b_1 + b_2$. The construction of the initial reference set starts with the selection of the best b_1 solutions from P. These solutions are added to *RefSet* and deleted from P. For each solution in $P \setminus RefSet$, the minimum of the distances to the solutions in *RefSet* is computed. Then, the solution with the maximum of these minimum distances is selected. This solution is added to *RefSet* and deleted from P and the minimum distances are updated. (In applying this max-min criterion, or any criterion based on distances, it can be important to scale the problem variables, to avoid a situation in which a particular variable or subset of variables dominates the distance measure and distorts the appropriate contribution of the vector components.) The process is repeated b_2 times, thus the resulting reference set has b_1 high quality solutions and b_2 diverse solutions.

The Linear Ordering Problem

In Chap. 2 we reviewed constructive methods for the LOP: G1, G2, G3, G4, G5, G6, MIX, RND, DG, and FQ. Any of them can be used as a diversification generation method within a SS algorithm. Moreover, we have seen that FQ, which is based on frequency information, performs best, considering quality and diversity as equally important. Therefore, we will employ this generator in our SS method for the LOP.

In the standard SS design [105] the improvement method is applied to all the solutions in P. However, in some problems where local search methods are usually extremely time consuming, the improvement method is not applied across the board but rather in a selective manner. In the LOP we follow the standard design and apply the improvement method to all the solutions in P. In Sect. 3.2 we describe a local search procedure using insertions. Based on moves $move(p_j, i)$, the local search method performs steps as long as the objective function increases. We apply this local search as the improvement method of our SS algorithm.

The Maximum Diversity Problem

Gallego et al. [61] implemented three variants of the scatter search procedure for the MDP. The differences among these implementations are related to the methods used to construct, combine and improve solutions, namely Base, GRASP Hybrid, and Tabu Search Hybrid. The Base method does not use any specific information about the problem context. The GRASP Hybrid implements several strategies that take advantage of characteristics that are specific to the MDP following the guidelines of this methodology. The Tabu Search Hybrid uses both context information and memory structures that are typical to tabu search implementations.

The Base procedure generates diversification by simply selecting m elements at random. The diversification generated in this way refers to the distances between the solutions and not about the objective function values. The GRASP Hybrid employs a more elaborate procedure for generating diverse solutions. The procedure, developed by Duarte and Martí [51], is based on randomizing the deterministic destructive heuristic D2 developed by Glover et al. [68]. The method, called GRASP-D2 is described in Section 3.2.1.

The diversification generator within the Tabu Search Hybrid variant is also based on the destructive procedure D2. As it is customary in tabu search, this method implements memory information instead of randomization. In particular, to deselect and element i from the partial solution in which more than m elements are selected, the method considers both its frequency $f(i)$ (number of previous solutions in which it was present) and its quality $q(i)$ (the average quality of the previous solutions in which it was present). With both values, the method computes the following expression $d(i)$:

$$d(i) = \alpha \frac{f(i)}{fmax} - \beta \frac{q(i)}{qmax} + \sum_{u \in M} d_{ui},$$

where $fmax$ and $qmax$ are the maximum of the frequencies and objective values respectively, and α and β their respective weights. At each step the algorithm deselects the element with minimum d-value. The method stops when there are exactly m elements selected.

Gallego et al. [61] coupled the three diversification methods above with three improvement methods respectively to populate Scatter Search with high-quality and, at the same time, diverse solutions. In particular they consider a so-called best improve local search, first improve local search, and short term memory tabu search. The best improve local search scans the set of selected elements in search of the best exchange to replace a selected element with an unselected one. The method performs moves as long as the objective value increases and it stops when no improving exchange can be found. This is similar to the improvement method proposed in [66]. The first improvement local search is the one applied in GRASP-D2 and mentioned above.

The short-term tabu search method begins with a random selection of an element i. The probability of selecting element i is inversely proportional to $d(i)$. The list of unselected elements is scanned and the first improving move that exchanges elements i and j is selected. If no improving move is found, then the least non-improving move is chosen. The chosen exchange is performed and both elements participating in the exchange are classified tabu-active for a number of iterations (known as the tabu tenure). The method stops if after a number of consecutive iterations the incumbent solution is not modified.

An interesting characteristic of this tabu search implementation consists in using an asymmetric tabu tenure in which elements added to the solution are given shorter tabu tenures than the tenure assigned to those elements that have been deleted from the solution. Also, the tabu tenure and the maximum number of iterations have been made dependent on the number of elements in the solution. According to the experimentation reported in [51], the tabu tenure for selected elements is set to 0.28m, while the tabu tenure for unselected elements is set to 0.028m.

3.7.2 Reference Set Update

The search is initiated in this second stage by applying the subset generation method that, in its simplest form, involves generating all pairs of reference solutions. The pairs of solutions in *RefSet* are selected one at a time and the solution combination method is applied to generate one or more trial solutions. These trial solutions are subjected to the improvement method. The reference set update method is applied once again to build the new *RefSet* with the best solutions, according to the objective function value, from the current *RefSet* and the set of trial solutions. The basic procedure terminates after all the subsets generated are subjected to the combination method and none of the improved trial solutions are admitted to *RefSet* under the rules of the reference set update method.

Solution combination methods in scatter search typically are not limited to combining just two solutions and therefore the subset generation method in its more general form consists of creating subsets of different sizes. The scatter search methodology is such that the set of combined solutions (i.e., the set of all combined solutions that the implementation intends to generate) may be produced in its entirety at the point where the subsets of reference solutions are created. Therefore, once a given subset is created, there is no merit in creating it again. This creates a situation that differs noticeably from those considered in the context of genetic algorithms, where the combinations are typically determined by the spin of a roulette wheel (see next section for a description).

The procedure for generating subsets of reference solutions in advanced SS applications uses a strategy to expand pairs into subsets of larger size while controlling the total number of subsets to be generated. In other words, the mechanism does not attempt to create all subsets of size 2, then all the subsets of size 3, and so on until reaching the subsets of size $b - 1$ and finally the entire *RefSet*. This approach would not be practical because there are 1023 subsets in a reference set of a typical size $b = 10$. Even for a smaller reference set, combining all possible subsets is not effective, because many subsets will be almost identical. For example, a subset of size four containing solutions 1, 2, 3, and 4 is almost the same as all the subsets with four solutions for which the first three solutions are solutions 1, 2 and 3. And even if the combination of subset $\{1, 2, 3, 5\}$ were to generate a different solution than the combination of subset $\{1, 2, 3, 6\}$, these new trial solutions would likely converge to the same local optimum after the application of the improvement method.

The following approach selects representative subsets of different sizes by creating subset types:

- **Type 1**: all 2-element subsets,
- **Type 2**: 3-element subsets derived from the 2-element subsets by augmenting each 2-element subset to include the best solution not in this subset,
- **Type 3**: 4-element subsets derived from the 3-element subsets by augmenting each 3-element subset to include the best solutions not in this subset, and
- **Type 4**: the subsets consisting of the best i elements, for $i = 5$ to b.

The Linear Ordering Problem

We will use this approach in our subset generation method for the LOP. In [24]
an experiment is designed with the goal of assessing the contribution of combining subset types 1 to 4 in the context of the LOP. The experiment tried to identify
how often, across a set of benchmark problems, the best solutions came from combinations of reference solution subsets of various sizes. Since subset types 1 to 4,
respectively, generate solutions from 2 to up to b reference solutions, it is sufficient
to keep a 4-element array for each solution generated during the search. The first
element of the array is the counter corresponding to subset type 1; the second element corresponds to subset type 2, etc. The array for each solution in the initial
reference set starts as (0,0,0,0), meaning that there are no sources. The array then
counts the number of times the different subset types are used. E.g., suppose that
three solutions in a subset of type 2 with arrays $[2,0,0,1]$, $[5,1,0,0]$ and $[0,1,0,0]$
are combined. Then a new solution resulting from this combination is accompanied
by the array $[7,3,0,1]$, the sum of the other arrays, plus 1 added to position 2.

In an experiment with 15 input-output instances from LOLIB tracking arrays
are used in [24] to find the percentage of times that each subset type produces solutions that become members of the reference set. The same experiment was also
conducted employing 15 randomly generated instances with 100 objects and with
weights drawn uniformly distributed between 0 and 100. The percentages are shown
in Figure 3.8, where the bars labeled "LOLIB" represent the results from the experiments with the input-output instances and the bars labelled "Random" correspond
to the results from the randomly generated instances.

As noted before, the combination method is an element of scatter search whose
design depends on the problem context. Although it is possible in some cases to
design context independent combination procedures, it is generally more effective
to base the design on specific characteristics of the problem setting.

The combination method for the LOP employs a min-max construction based
on votes. The method scans (from left to right) each solution in a subset (called
the reference permutation), and uses the rule that each reference permutation in the
combination subset votes for its first object that so far has not been included in the
combined permutation (referred to as the "incipient object"). The voting determines
the object to be assigned to the next free position in the combined permutation
(where the incipient object with more votes is assigned). This is a min-max rule
in the sense that, if any object of the reference permutation is chosen other than
the incipient object, then it would increase the deviation between the reference and
the combined permutations. Similarly, if the incipient object were placed later in
the combined permutation than its next available position, this deviation would also
increase. So the rule attempts to minimize the maximum deviation of the combined
solution from the reference solution, subject to the fact that other reference solutions
in the subset are also competing to contribute.

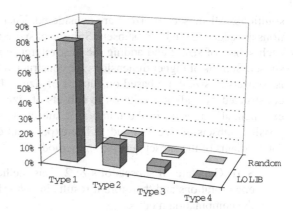

Fig. 3.8 Performance of
subset types in the LOP

This voting scheme can be implemented using a couple of variations that depend
on the way votes are modified:

– The vote of a given reference solution is weighted according to the incipient
  object's position (referred to as the "incipient position"). A smaller incipient
  position gets a higher vote. For example, if the object in the first position of
  some reference permutation is not assigned to the combined permutation during the first 4 assignments, then the vote is weighted more heavily to increase
  the chances of having that object assigned to position 5 of the combined permutation. The rule emphasizes the preference of this assignment to one that
  introduces an (incipient) object that occurs later in some other reference permutation.
– A bias factor gives more weight to the vote of a reference permutation with
  higher quality. Within the current organization of the scatter search implementation in this tutorial such a factor should only have weak influence because it
  is expected that high quality solutions will be strongly represented anyway.

We chose to implement the first variant with a tie-breaking rule based on solution
quality. The tie-breaking rule is used when more than one object receives the same
votes. Then the object with highest weighted vote is selected, where the weight of
a vote is directly proportional to the objective function value of the corresponding
reference solution.

The Maximum Diversity Problem

The subset generation method for the MDP implements the straight-forward design
described above, in which all pairs of reference solutions are generated. Then, they
are selected one at a time, and the solution combination method is applied to generate one or more trial solutions.

   Gallego et al. [61] performed an experiment to identify the most effective method
for generating subsets of reference solutions that are in turn the input to the combination method. For this experiment, they consider combinations of 2, 3, 4 and 5

solutions. All subsets of size 2 are considered. That is, all pairs of reference solutions are added to the list of subsets. Subsets of size 3 are constructed by considering each subset of size 2 and adding the best reference solution that is not part of the subset. Subsets of higher dimensions are constructed following the same logic. That is, subsets of size 4 are based on subsets of size 3. Likewise, subsets of size 5 are constructed by adding a solution to subsets of size 4. This mechanism avoids the exponential explosion in the number of subsets generated had we considered all possible subsets of size 3, 4 and 5. The experiment consists of testing the merit of four variants of the subset generation method:

- SG1: Generate all subsets of size 2. This method generates all pairs of reference solutions and therefore it results in $b(b-1)/2$ subsets that are passed to the combination method.
- SG2: Generate all subsets of size 2 and then augment each pair to generate subsets of size 3. The way a solution is added to each pair creates duplicates and therefore $b(b-1)$ is an upper bound on the number of subsets generated by this variant.
- SG3: Augment SG2 with subsets of size 4 that are generated by adding a solution to the subsets of size 3.
- SG4: Augment SG3 with subsets of size 5 that are generated by adding a solution to the subsets of size 4.

The results of running this experiment are summarized in table 3.10 reporting the average deviation with respect to the best known solution, and the number of instances in which the method obtains the best solution over 20 representative instances.

Table 3.10 Subset generation designs for the MDP

Subset Generation	Average Deviation	Number of Best
SG1	0.0000%	20
SG2	0.0017%	19
SG3	0.0000%	20
SG4	0.0042%	18

The results of this experiment indicate that there is no additional gain that could be realized by generating and combining subsets with more than 2 solutions. Hence, the scatter search implementation for the MDP limits the subset generation to all pairs of reference solutions.

In line with the three implementations for the diversification generation method described above, Gallego et al. [61] proposed three ways to combine two solutions: random, greedy, and memory-based. The random selection simply consists in selecting m elements at random from the union of the elements in both reference solutions to be combined. The greedy randomized variant has two steps as well, in the first one it computes the union of the solutions being combined. In the second step, it removes from the union the elements with lower contribution. Specifically, at each step the algorithm deselects the element with minimum sum of distances to the rest of selected elements, until only m elements remain selected. Finally, the memory-based implementation is similar to the greedy one, but the sum of distances used to deselect elements from the union, is coupled with the frequency and objective function information, thus computing the d-values described above.

3.7.3 Reference Set Rebuild

In basic scatter search implementations, the reference set is updated by replacing the reference solution with the worst objective function value with a new trial solution that has a better objective function value. Since we always assume that *RefSet* is ordered, the best solution is x_1 and the worst solution is x_b. So, when a new trial solution x is generated as a result of the application of the combination and improvement methods, the objective function value of the new trial solution x is used to determine whether *RefSet* needs to be updated (it is updated by replacing x_b with x and reordering its elements). If the reference set remains unchanged after the update (no trial solution improves the worst solution x_b in the set), then a rebuild step is performed.

A reference set $\{x_1, x_2, \ldots, x_{b_1}, x_{b_1+1}, \ldots, x_{b_1+b_2}\}$, where $b = b_1 + b_2$, is partially rebuilt with the following diversification update when no new trial solutions are admitted to it:

(1) Solutions $x_{b_1+1}, \ldots, x_{b_1+b_2}$ are deleted from *RefSet*.
(2) The diversification generation method is used to construct a set P of new solutions.
(3) We sequentially select b_2 solutions from P and move them to *RefSet*. We apply the same min-max criterion, which is part of the reference set update method, as we did when *RefSet* was constructed the first time.

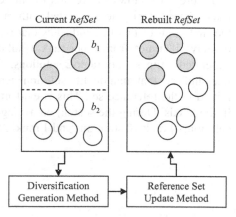

Fig. 3.9 *RefSet* rebuilding

Figure 3.9 shows a schematic representation of the rebuilding mechanism. The rationale behind it is that, when the search is stagnated because the solutions in the RefSet are similar, we input diversity in the RefSet by generating new and diverse solutions.

The Linear ordering Problem

A series of experiments is performed to assess the quality of the SS implementation in [24] for the LOP. We report here on three of them. The first one is designed to find the best values for the key parameters of the scatter search algorithm. For this experiment, again 15 input-output instances and 15 randomly generated instances with 100 objects (as described above) are used.

The following values were tested during these experiments:

- $PSize$: 50, 100, 150,
- b: 10, 20, 40,
- (b_1, b_2): $(5,5)$, $(10,10)$, $(5,15)$, $(15,5)$, $(20,20)$.

The experiments revealed that a significant change in the solution quality is due to the increase in $PSize$. The experiments were inconclusive about the advantage of increasing the size of the reference set (i.e., the value of b) beyond 20 when $PSize$ is not greater than 100. The experiments showed that the best results are obtained when $b_1 = b_2$. Therefore the key parameters were set as $PSize = 100$, $b = 20$ and $b_1 = b_2 = 10$.

The second experiment was performed to learn about the ranks of the reference solutions that generated the best solutions found in the search process. To this end, it was tracked from which rank or position in the reference set the best solutions came from. Say e.g., the overall best solution came from combining the 3rd and 5th best solutions, where one of these came from combining the 1st, 2nd and 6th best solutions, and the other came from ... etc. This trace would give an idea of which solutions are important as components of others.

Figure 3.10 shows the results of this experiment. Its interpretation is as follows. Consider the line associated with rank 1. Then, the count of (almost) 18 in the first point of this line indicates that rank 1 solutions were generated (approximately) 18 times from other rank 1 solutions. Similarly, rank 1 solutions were generated 8 times from rank 2 solutions. The decaying effect exhibited by all the lines indicate that high quality solutions tend to generate new solutions that are admitted to the reference set. This is evident by the counts corresponding to rank 1 in the x-axis. During the search, on average, rank 1 solutions generated 18 rank 1 solutions, 4 rank 10 solutions and 1 rank 20 solution.

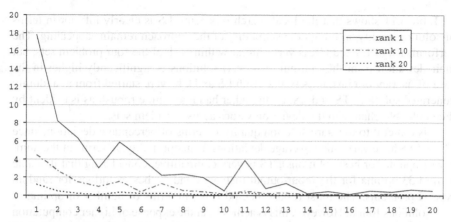

Fig. 3.10 Source ranks of reference solutions

In the third experiment the SS algorithm is compared with some of the state-of-the-art meta-heuristics for the LOP on the input-output matrices, SGB instances and random problems A (types I and II). In particular, we report about a comparison of SS (with its parameters set as specified above) with the method CK of [31] and the improved GRASP method LS (described in Sect. 3.2). Again, CK-10 denotes the application of CK from 10 randomly generated initial solutions.

Table 3.11 shows, for each method, the average percent deviation from optimality (or from the best known solutions), the number of optimum (or best) solutions, and the average CPU time (seconds on a Pentium 166 MHz).

Table 3.11 Comparison of best methods

	LS	CK	CK-10	TS	SS
Input-output instances					
Deviation	0.15	0.15	0.02	0.04	0.01
No. of opt. solutions	11	11	27	33	42
CPU time	0.01	0.10	1.06	0.49	2.35
SGB instances					
Deviation	0.31	0.05	0.01	0.01	0.01
No. of opt. solutions	0	0	0	5	18
CPU time	0.10	2.73	31.93	3.03	14.05
Random A type I					
Deviation	0.45	0.48	0.29	0.11	0.02
No. of best solutions	0	0	0	5	21
CPU time	0.08	6.90	67.12	12.74	63.82
Random A type II					
Deviation	0.02	0.02	0.01	0.00	0.00
No. of best solutions	0	0	0	41	14
CPU time	0.07	4.30	44.51	7.91	43.36

Table 3.11 shows that the local search procedure LS is clearly inferior in terms of solution quality, although the simplicity of the approach remains appealing. The performance of LS and CK is very similar within each of the four problem sets, but their deviation from the optimum (or best) solutions is significantly higher in the case of the random instances type I. Both LS and CK were started from a randomly generated solution. TS and SS, on the other hand, are quite robust, as is evident by the negligible change in the deviation values across problem sets.

It is difficult to measure solution quality in terms of percentage deviation, since TS and SS have very small average deviations from optimality. In terms of the number of optima (or best solutions), TS is very competitive, considering that it is able to find 33 optima for the input-output instances and 41 best solutions for random type II problems. The most robust method is SS in terms of number of optima or best solutions found. However, this is achieved at the expense of higher computation times.

Finally, we report the results obtained when applying SS for 10 seconds to the 229 instances in the OPT-I set. Table 3.12 reports the #Instances, *Dev* and *#Opt* statistics in each subset of instances.

Table 3.12 Scatter Search on OPT-I instances

	IO	SGB	RandAII	Rand B	MB	Special	**Total**
#Instances	50	25	25	70	30	29	229
Dev(%)	0.01	0.00	0.01	0.01	0.00	0.09	0.02
#Opt	42	20	13	62	16	12	165

The results in Table 3.12 clearly show that the Scatter Search methodology obtains high quality solutions on the OPT-I set. The average percentaged deviation overall is 0.02% and the number of optimum solutions is 165 out of 229 instances.

The Maximum Diversity Problem

Gallego et al. [61] performed an interesting experiment to disclose the best size of the RefSet, usually called b. Figure 3.11 shows the average percent deviation obtained for the different b-values over a collection of representative instances from the MDPLIB.

The diagram depicted in Figure 3.11 shows that the best results are found for $b = 12, 20$ and 40, with 0.000% average percent deviation, and low quality results are obtained with extreme b-values (i.e., smaller than 8 or larger than 60). It must be noted that this result confirms the usual recommendation in Scatter search tutorials of setting b with a relatively low value close to 10.

An important difference between genetic algorithms and scatter search is that the latter iterates over the RefSet containing a few good solutions, while genetic algorithms iterate over a relatively large set of solutions called the population, which usually contains about 100 solutions. Additionally, scatter search performs an exhaustive exploration of the RefSet, usually combining all its pairs of solutions, while genetic algorithms apply statistical sampling to explore the population. In the next section we will examine genetic algorithms.

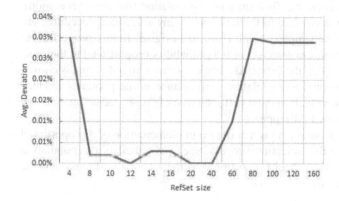

Fig. 3.11 RefSet size versus quality

3.8 Genetic and Memetic Algorithms

The idea of applying the biological principle of natural evolution to artificial systems, introduced more than three decades ago, has seen impressive growth in the past few years [125].

Usually grouped under the term *evolutionary algorithms* or *evolutionary computation*, we find the domains of *genetic algorithms, evolution strategies, evolutionary programming*, and *genetic programming*. Evolutionary algorithms have been successfully applied to numerous problems from different domains, including optimization, automatic programming, machine learning, economics, ecology, population genetics, studies of evolution and learning, and social systems.

A genetic algorithm is an iterative procedure that consists of a constant size population of individuals, each one represented by a finite string of symbols, known as the genome, encoding a possible solution in a given problem space. This space, referred to as the search space, comprises all possible solutions to the problem at hand. Generally speaking, genetic algorithms are applied to spaces that are too large to be exhaustively searched (such as those in combinatorial optimization). Solutions to a problem were originally encoded as binary strings due to certain computational advantages associated with such encoding. Also the theory about the behavior of algorithms was based on binary strings. Because in many instances it is impractical to represent solutions using binary strings, the solution representation has been extended in recent years to include character based encoding, real-valued encoding, and tree representations.

The standard genetic algorithm [44] proceeds as follows: an initial population of individuals is generated at random or heuristically. In every evolutionary step, denoted as a generation, the individuals in the current population are decoded and evaluated according to some predefined quality criterion, referred to as their fitness (evaluated by a fitness function). To form a new population (the next generation), individuals are selected according to their fitness. Many selection procedures are currently in use, one of the simplest being Holland's original fitness-proportionate selection, where individuals are selected with a probability proportional to their relative fitness. This ensures that the expected number of times an individual is chosen is approximately proportional to its relative performance in the population. Thus, high fitness ("good") individuals stand a better chance of "reproducing", while low fitness ones are more likely to disappear.

The *roulette wheel selection* [41] is a common implementation of a proportional selection mechanism. In this selection process, each individual in the population is assigned a portion of the wheel proportional to the ratio of its fitness and the population's average fitness.

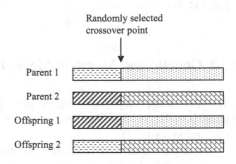

Fig. 3.12 One point crossover

Tournament and *ranking* are two other popular selection techniques. Tournament selection consists of choosing q individuals from a population of n individuals and selecting the best according to the fitness value to survive into the next generation. Hence, n tournaments are necessary to build the population for the next generation. Binary tournaments, for which $q = 2$, are the most common implementation of this selection technique. Ranking ignores the fitness values and assigns selection probabilities based exclusively on rank. Genetically inspired operators are used to introduce new individuals into the population, i.e., to generate new points in the search space. The best known such operators are *crossover* and *mutation*. Crossover is performed, with a given probability (the crossover probability or crossover rate), between two selected individuals, called parents, by exchanging parts of their genomes (i.e., encoding) to form two new individuals, called offspring; in its simplest form, substrings are exchanged after a randomly selected crossover point. This operator tends to enable the evolutionary process to move toward promising regions of the search space. Figure 3.12 depicts a one point crossover operation.

The mutation operator is introduced to prevent premature convergence to local optima by randomly sampling new points in the search space. Mutation entails flipping bits at random, with some (small) probability. Figure 3.13 shows a graphical representation of the mutation operator.

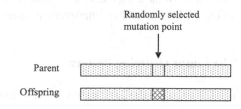

Fig. 3.13 Mutation operation

The principle layout of a genetic algorithm is the following.

Genetic algorithm

(1) Create the initial population P by randomly generating n solutions.

(2) While the stopping condition is not met:

    (2.1) Evaluate the solutions in P and update the best solution found if necessary. (*Evaluate*)

    (2.2) Calculate the probability of surviving based on solution quality. Evolve P by choosing n solutions according to their probability of surviving. (*Survival of the fittest*)

    (2.3) Select a fraction pc of the solutions in P to be combined. Selection is at random with the same probability for each element of P. The selected elements are randomly paired for combination, with each pair generating one or more offspring that are added to P. (*Combine*)

    (2.4) A fraction pm of the solutions in P is selected for mutation. The mutated solution is improved and added to P. (*Mutate*)

(3) Output the best solution in P.

Genetic algorithms are stochastic iterative processes that are not guaranteed to converge in practice; the termination condition may be specified as some fixed, maximal number of generations or as the attainment of an acceptable fitness level for the best individual. However, contrary to other meta-heuristics, theoretical studies [44], based on Markov chains and the so-called *schema theorem*, have been developed to establish the convergence conditions to a global optimum with respect to the selection strategies and operators of the method.

In general terms, we may say that *Memetic Algorithms* (MAs) constitute a class of solving methods with a particular structure given by population, generational evolution, and local search. Neri and Cotta [131] characterize them as a flexible class of algorithms, containing the previous evolutionary algorithms, and combining global and local search. In a basic MA, the initial population is generated following a systematic procedure. Then, the method iterates over a main loop that basically consists of three elements: cooperate, improve, and compete. Cooperate and improve constitute the core of the MA, and diversity management is one of the key points in the interaction of these elements. In simple terms, we may say that to create a MA from a GA, we have to add an improvement method.

The Linear Ordering Problem

In [32] a GA for the LOP based on the following three elements is proposed:

(1) A selection procedure. The n individuals in the current population are sorted according to their objective value (where the best one comes first). Then, the probability to choose an individual for combination is proportional to its rank in this ordering (i.e., the best individual has a selection probability n times the probability of the worst one).

(2) A crossover operator. Two selected solutions (individuals) are "crossed" or combined to obtain a new one. Specifically, given $u = \langle u_1, u_2, \ldots, u_n \rangle$ and $v = \langle v_1, v_2, \ldots, v_n \rangle$, we first compute the auxiliary array $a = (a_1, a_2, \ldots, a_n)$ where $a_i = u_i + v_i$. Then, we order the elements in a from the lowest to the highest: $a_{\sigma(1)} \leq a_{\sigma(2)} \leq \ldots \leq a_{\sigma(n)}$. The offspring w is constructed as

$$w = \langle a_{\sigma(1)}, a_{\sigma(2)}, \ldots, a_{\sigma(n)} \rangle.$$

(3) A mutation operator. Regardless of the objective function, a selected individual is mutated by applying a simple transformation.

Table 3.13 reports the results obtained with our implementation of a classical GA for the LOP according to the three elements above. This table shows the #Instances, *Dev* and *#Opt* statistics in each subset of instances in the OPT-I set. As in our previous experiments, we run the method for 10 seconds on each instance.

Table 3.13 Genetic Algorithm on OPT-I instances

	IO	SGB	RandAII	Rand B	MB	Special	**Total**
#Instances	50	25	25	70	30	29	229
Dev(%)	0.38	0.76	21.44	0.75	35.53	1.38	10.04
#Opt	9	0	0	1	0	4	14

The results in Table 3.13 show that the GA obtains relatively low quality results on the OPT-I instances. It is only able to obtain 14 optimum solutions out of 229 instances. Note that the methods reported in the previous sections obtain a larger number of optimum solutions (215 in the case of the tabu search).

In [154] a genetic algorithm coupled with a local search procedure is developed for the LOP. This hybrid method is called *memetic algorithm*. In the initialization, a population of individuals is obtained by first generating a set of random permutations (solutions) and then applying a local search procedure to each of them. The local procedure is based on insertions where the neighborhood is examined in random order and the first improving move is performed. This method is very similar to the local search procedure implemented for the LOP in other meta-heuristics described in previous sections.

In each iteration of the algorithm, called *generation*, new solutions are generated by applying crossover and mutation to randomly selected solutions in the population (according to a uniform distribution). The crossover operator takes two individuals of the current population and combines them into a new individual, while the mutation operator introduces a perturbation into an individual. Local search is applied again to improve each new solution. The new population is created by merging the best solutions in the population and the new improved solutions. It is worth mentioning that the authors consider four different crossover operators: DPX (similar distance from parents), CX (classical crossover), OB (order based crossover) and Rank (computing the average ranking of the elements). In computational experiments CX and OB performed best.

Finally, if the average of the objective function value of the population does not change in a certain number of consecutive generations, the population is rebuilt in a similar way as the Reference Set is rebuilt in scatter search: the best solutions are kept and the worst solutions are replaced with new ones.

Table 3.14 shows the results obtained with the memetic algorithm [154] on the OPT-I instances. We run the method for 10 seconds and report the #Instances, *Dev* and *#Opt* statistics in each subset of instances.

Table 3.14 Memetic Algorithm on OPT-I instances

	IO	SGB	RandAII	Rand B	MB	Special	**Total**
#Instances	50	25	25	70	30	29	229
Dev(%)	0.00	0.00	0.00	0.00	0.00	0.00	0.00
#Opt	50	25	25	70	30	28	228

Table 3.14 clearly shows that the MA obtains the best results so far on the OPT-I instances (it only misses one optimum solution out of 229 instances). Comparing these results with those obtained with a classical GA method (reported in Table 3.13) we can conclude that the inclusion of a local search makes an important difference in a meta-heuristic procedure.

In [89] a similar method, based on combining a classical GA with a local search is presented. Is is called *hybrid genetic algorithm* and it is very similar to the method in [154]. The local search is also based on exchanges and it also applies the CX and OB crossover operators. However, instead of DPX and Rank, it applies PMX (partially matched crossover). In their experimentation the authors conclude that GA without a local search produces low quality results, but its hybridization with local search is able to match state-of-the-art methods.

An interesting variant is proposed by Garcia et al. [63] in a hybrid method called Hybrid Exploration Algorithm (HEA) that reduces the scope of the exploration as the method evolves. In particular, once the population has been initialized, HEA iterates on the next instructions. Firstly, the individuals of the population are moved (perturbed) with a certain number of swaps, and improved with a local search (under the insert neighborhood) to create an auxiliary population. Secondly, the best solutions of the newly created population and the former one are mixed, and the best elements are chosen, while the rest of the solutions are discarded. When a solution is perturbed considering a high number of swaps, it is expected that the obtained solution is far from the one perturbed. On the contrary, a low number of swaps results in a similar solution. Assuming this principle, at the beginning of the optimization, HEA perturbs the solutions in the population with a high number of swaps in order to locate the regions with the best solutions in the search space, and, later, progressively decreases the number of swaps to explore in detail the chosen regions. The number of random swaps starts with $0.45n$ and decreases at a rate of 5% each time until $0.05n$.

We can establish that both scatter search and genetic algorithms belong to the family of population based meta-heuristics. Moreover, they were both proposed in the seventies: while Holland [87] introduced genetic algorithms and the notion of imitating nature and the "survival of the fittest" paradigm, Glover [70] introduced scatter search as a heuristic for integer programming that expanded on the concept of surrogate constraints. Both methods are based on maintaining and evolving a population of solutions throughout the search. Although the population based approach makes SS and GA similar in nature, as described in [105], there are fundamental differences between these two methodologies:

-   The population in genetic algorithms is about one order of magnitude larger than the reference set in scatter search. A typical GA population size is 100, while a typical SS reference set size is 10.
-   A probabilistic procedure is used to select parents to apply crossover and mutation operators in GAs while the combination method is applied to a predetermined list of subsets of reference solutions in scatter search.
-   The evolution of a GA population follows the "survival of the fittest" philosophy, which is implemented using probabilistic rules. In scatter search, changes in the reference set are controlled by the deterministic rules in the reference set update method.
-   The use of local search procedures is an integral part of scatter search, while it was added to GAs in order to create hybrid implementations that would yield improved outcomes.
-   The subset generation method considers combinations of more than two solutions. GAs are typically restricted to combining two solutions.
-   Full randomization is the typical mechanism used in GAs to build the initial population. Diversity in scatter search is achieved with the application of the

diversification generation method, which is designed to balance diversity with solution quality.

Recently, Ceberio et al. [29] study the influence that the positions of the indexes that compound a solution in the permutation have when generating local optimal solutions. As a result of their theoretical study, a restricted version of the insertion neighbourhood is proposed. This neighbourhood discards specific insertion moves that involve moving indexes to positions at which they cannot generate local optimal solutions. The theoretical analysis proves that these insertion operations would never improve the most a solution compared with other neighborhood solutions, and therefore can be discarded in the neighbourhood exploration. Considering that restricted positions would not generate local optimal solutions, they can be discarded in the search for the global optimum.

The authors also perform an empirical analysis of their restricted neighborhood, by implementing it in two of the best previous methods, the MA described above, and the Iterated Local Search [154]. Experimental results show that their restricted variants outperform the standard designs in 90 percent of the cases tested, obtaining the same results for the rest of them. Moreover, the experiments devoted to measure the execution time show that the restricted approaches are, as expected, faster than the classical versions in most of the instances. The paper concludes pointing out possible extensions of this work that deserve special attention. In particular, taking into account that the restrictions matrix describes partially the global optimum solution, such information could be used in exact, heuristic, and metaheuristic algorithms in order to guide the search toward better solutions. It is indeed very interesting to include the restrictions matrix into Branch and Bound algorithms to discard the branches that do not comply with the restrictions, and test how it affects the performance of the method. This work clearly opens new possibilities to designs advanced solving methods.

The Maximum Diversity Problem

Zhou et al. [165] implemented a typical MA algorithm for the MDP. In a high level description, we may say that the method begins with a set of constructed solutions (initial population). At each generation, MA selects two or more parent solutions from the population, and performs a recombination or crossover operation to generate one or more offspring solutions. Then a local optimization procedure is invoked to improve offspring solutions. Finally, a population management strategy is applied to decide if each improved offspring solution is accepted to join the population. The process repeats until a stopping condition is satisfied. The authors hybridize MA with an advanced strategy taken from machine learning. In particular, they consider opposition based learning as a basis to improve the standard MA, creating what they called *Opposition Based Memetic Algorithm* (OBMA).

A more detailed description of the method follows. OBMA starts from a collection of diverse elite solutions which are obtained by the opposition-based initialization procedure. At each generation, two parent solutions are selected at random from the population, and then the backbone-based crossover operator is applied to the selected parents to generate an offspring solution and a corresponding opposite solution. Then, the opposition-based search procedure is applied to search from the offspring solution and its opposite solution. Finally, the rank-based pool updating strategy decides whether these two improved offspring solutions should be inserted into the population. This process repeats until the time limit is reached.

The main strategy in opposition based learning is the simultaneous consideration of a solution and its corresponding opposite solution. Given M a feasible solution, and x its associated binary vector, an opposite solution S corresponds to a feasible binary vector whose components match the complement of x as closely as possible (selecting one at random among the closest). The complement of a binary vector x is the vector y with the same number of components verifying that $y(i) = 1$ if $x(i) = 0$, and $y(i) = 0$ if $x(i) = 1$.

The opposition-based initialization procedure generates pairs of solutions. Specifically, one at random and a corresponding opposite solution. These two solutions are then improved by a tabu search procedure, and the best improved solution is inserted into the population under construction.

The backbone-based crossover operator is similar to the combination method in Gallego et al. [61]. In particular, given two solutions, the method first selects their common elements, and then completes them with additional elements to reach the required number of selected elements m. There is an important difference here between both methods. While the scatter search combination method in [61] evaluates a greedy function based on frequency and quality to select the additional elements from the entire set of unselected ones, the crossover only considers the elements in the two parent solutions. Once the offspring solution is obtained, the method generates the opposite solution, thus obtaining two solutions with the application of the crossover operator.

To maintain a good level of diversity in the population and avoid a premature convergence, the memetic algorithm applies a rank-based pool updating strategy to decide whether the improved solutions are inserted into the population. This pool updating strategy simultaneously considers solution quality and inter-distance between solutions to keep population diversity.

It is interesting to observe that in the parameter setting, Zhou et al. [165] determine that the best size for the population is 10, which is in line of what Gallego et al. [61] found for their scatter search. This is to be expected, since scatter search can be considered itself a memetic algorithm with specific characteristics. Nevertheless, there are still important difference between both methods, specially in the way solutions are selected from the population for combination. On the other hand, the improvement method in [165] contains a tabu search algorithm, making it a memory-based metaheuristic. According to the experimentation, this method presents state-of-the art performance, achieving the best results reported so far.

Although GA and SS have contrasting views about searching a solution space, it is possible to create a hybrid approach without entirely compromising the scatter search framework. Specifically, if we view the crossover and mutation operators as an instance of a combination method, it is then straightforward to design a scatter search procedure that employs genetic operators for combination purposes.

3.9 Matheuristics

The term *matheuristic* refers to the hybridization of mathematical programming with heuristics. They are usually implemented in a bi-level scheme, which is also called in Computer Science a master-slave framework. In some applications, the heuristic controls the main flow of the solving method, calling repeatedly the exact method to solve a subproblem of the original problem (either in size or scope). Alternatively, the exact method, or the mathematical formulation, can be the one playing the role of the master, calling the heuristic, or several heuristics, at different stages of the solving process. In the metaheuristic domain, the former scheme is the most common, and most of the matheuristic consists in sequentially applying a mathematical programming solver, such as CPLEX or Gurobi, as the main element in the heuristic process.

Garcia et al. [63] proposed the Sequential Exact Improvement (SEI algorithm) for the LOP, based on the exact resolution of a reduced subproblem with its binary formulation. Considering that the contribution of an element to the objective function is independent to the ordering of the previous and posterior elements in the solution, it is reasonable to design strategies that optimize the ordering of subset of elements.

Based on the property above, and to speed up the solving process, SEI is applied to a relatively good solution, that can be obtained with any heuristic; in particular with the HEA described in the previous section. From this inital solution, the method sequentially solves sub-instances of the original instance, in which there is an overlapping of some elements. Specifically, half of the elements in a sub-instance have been included in the previously solved sub-instance, achieving in this way stability across iterations.

From the empirical experience of the ability of the binary formulation to solve instances in practical running times, SEI considers sub-instances of size 75. Experiments with instances with 150 and 250 elements show the effectiveness of the method, since it is able to identify new best solutions for most of the instances tested, although no information about their associated running time is provided.

3.10 Experiments with the LOP

In this section we compare the meta-heuristics described in the previous sections for the LOP. Specifically, we consider the following methods:

–   TS: Tabu Search

- MA: Memetic Algorithm
- VNS: Variable Neighbourhood Search
- SA: Simulated Annealing
- SS: Scatter Search
- GRASP: Greedy ramdomized adaptive search procedure
- GA: Genetic Algorithm

We divide our experimentation into two parts according to the classification of the instances introduced in Chap. 1. Table 3.15 reports the results on the 229 OPT-I instances and Table 3.16 reports those on the 255 UB-I instances.

Table 3.15 Meta-heuristics on OPT-I instances

	TS	MA	VNS	SA	SS	GRASP	GA
IO							
Dev(%)	0.00	0.00	0.00	0.03	0.01	0.00	0.38
Score	50	50	50	270	90	57	343
#Opt	50	50	50	16	42	49	9
SGB							
Dev(%)	0.00	0.00	0.00	0.03	0.00	0.00	0.76
Score	25	25	25	197	47	31	220
#Opt	25	25	25	0	20	23	0
RandAII							
Dev(%)	0.00	0.00	0.00	0.08	0.01	0.01	21.44
Score	25	25	47	197	80	116	242
#Opt	25	25	19	0	13	5	0
RandB 70							
Dev(%)	0.00	0.00	0.02	0.25	0.01	0.00	0.75
Score	70	70	103	486	115	70	607
#Opt	70	70	64	10	62	70	1
MB							
Dev(%)	0.00	0.00	0.00	1.33	0.00	0.00	35.53
Score	30	30	30	240	98	71	295
#Opt	30	30	30	0	16	21	0
Special							
Dev(%)	0.02	0.00	0.11	0.39	0.09	0.05	1.38
Score	64	30	66	166	109	84	226
#Opt	15	28	20	6	12	14	4
OPT-I							
Dev(%)	0.00	0.00	0.02	0.35	0.02	0.01	10.04
#Opt	215	228	208	32	165	182	14

In each experiment we compute for each instance and each method the relative deviation *Dev* (in percent) between the best solution value obtained with the method and the optimal value for that instance (in the UB-I instances we do not know the optimal value and therefore we instead consider the best known value). For each method, we also report the number of instances *#Opt* for which an optimum solution could be found (*#Best* in the case of UB-I instances). In addition, we calculate the so-called *score statistic* [147] associated with each method. For each instance, the *nrank* of method M is defined as the number of methods that found a better solution than the one found by M. In the event of ties, the methods receive the same *nrank*, equal to the number of methods strictly better than all of them. The value of *Score* is the sum of the *nrank* values for all the instances in the experiment. Thus the lower the *Score* the better the method.

Table 3.15 shows that most of the meta-heuristics considered are able to obtain all the optimal solutions within the time limit of 10 seconds considered (they actually obtain it in around 1 second). We therefore conclude that instances in OPT-I are easy for the best meta-heuristics and therefore not adequate to compare them.

In our second experiment we target the UBI-instances for which the optimum is not known but we have an upper bound for comparison. We therefore compute for each instance and each method the relative deviation *D.Best* (in percent) between the best solution value *Value* obtained with the method and the best known value *BestValue* as well as the relative deviation *D.UB* (in percent) between *Value* and the upper bound. For each method, we also report the number of instances *#Best* for which the value of the solution is equal to *BestValue*. As in the previous experiment we calculate the score statistic. Table 3.16 reports the values of these four statistics on the UB-I instances when running the 7 meta-heuristics for 10 seconds.

Table 3.16 Meta-heuristics on UB-I instances

	TS	MA	VNS	SA	SS	GRASP	GA
RandAI							
D.Best	0.12	0.05	0.47	1.77	0.26	0.42	10.59
D.UB	17.81	17.75	18.10	18.88	17.92	18.05	26.28
Score	201	105	482	641	326	461	936
#Best	5	33	0	0	1	0	0
RandAII							
D.Best	0.01	0.00	0.01	0.07	0.02	0.04	35.97
D.UB	0.38	0.38	0.39	0.44	0.40	0.41	36.21
Score	63	25	74	175	109	151	191
#Best	3	39	8	0	0	0	0
RandB							
D.Best	0.00	0.00	0.00	0.31	0.04	0.00	0.91
D.UB	3.20	3.20	3.26	3.51	3.24	3.20	4.08
Score	20	20	67	160	50	20	175
#Best	20	20	11	0	11	20	0
XLOLIB							
D.Best	0.62	0.12	0.42	0.53	0.68	1.14	23.99
D.UB	3.21	2.72	3.01	3.13	3.27	3.72	25.96
Score	307	87	200	266	320	460	758
#Best	0	2	0	0	0	0	0
Special							
D.Best	0.43	0.03	0.50	2.05	0.32	0.65	9.27
D.UB	9.61	9.26	9.67	11.04	9.52	9.81	17.35
Score	21	7	27	57	17	28	63
#Best	3	4	2	0	3	3	0
UB-I							
D.Best	0.23	0.04	0.28	0.95	0.26	0.45	16.15
D.UB	6.84	6.66	6.89	7.40	6.87	7.04	21.98
#Best	31	98	21	0	17	23	0

The results in Table 3.16 show that MA is able to obtain the largest number of best solutions (98 of a total of 255 instances) in short runs (10 seconds). No other method is able to obtain more than 31 best solutions, which clearly indicates the superiority of MA. On the other hand, considering average percentage deviations with respect to the best solutions, the differences among the methods appear to be very small. MA presents on average a deviation of 0.04% while TS, SS and VNS present averages deviations of 0.23% 0.26% and 0.28%, respectively. This indicates that although these methods are not able to match the best solution values, they obtain solutions with values very close to the best.

According to the differences among methods observed in Table 3.16, where the deviations w.r.t. the best solution known range from 0.04% to 17.72%, we can conclude that the instances in set UB-I are more difficult to solve than those in OPT-I (where the deviations range from 0.00% to 15.31%).

3.11 Experiments with the MDP

Two extensive empirical comparisons of the MDP have been published so far. In 2013, Martí et al. [116] compared all the methods published at that time, which includes 10 heuristics and 20 metaheuristics. This comparison revealed that simple heuristics, such as C2 and D2, perform very well considering their simplicity, and in the set of complex metaheuristics, B-VNS [20] and ITS [137] exhibit the best results. Since then, several new methods have been published, being the Memetic Tabu Search TS-MA [162] and the Opposition-based memetic algorithm OBMA [165] the best ones. Martí et al. [119] published this year 2021 an updated comparison, in which they extended the previous comparison by including these new methods. We summarize here their experimentation with the public domain set of instances MD-PLIB.

In line with previous comparisons, two time horizons are considering: 10 seconds and 600 seconds of CPU time. Table 3.17 reports the results of the six heuristics referenced above run for 10 seconds. It also reports the solutions of the CPLEX solver with the mathematical model run for 1 hour. Note that in many cases CPLEX is not able to certify the optimality, and we report its best feasible solution found (current lower bound when the time limit expires). This table shows the average percentage deviation from the best solution known (% *dev*), and the number of best solutions found (# *best*). Results are reported for each instance set. In the case of CPLEX, % *dev* is only reported in a set, when it obtains feasible solutions in all the instances in that set.

Table 3.17 shows that, as expected, metaheuristics obtain better results than simple heuristics. In particular, the most recent published method, OBMA, obtains the best results overall, with an average percentage deviation of 0.16% and 327 best solutions found in the experiment. Note that TS-MA is able to slightly improve OBMA in terms of the average percentage deviation; however, a *p*-value < 0.001 of the one-sided pairwise Wilcoxon test confirms the superiority of OBMA. On the other hand, this table also shows that most of the problems are too large to be solved with CPLEX, and only in some of the instances sets it obtains feasible solutions. It is interesting to see that the best nowadays methods are based on hybrid procedures that combine elements from different methodologies. in particular, TS-MA hybridizes tabu search with memetic algorithms, while OBMA is mainly based on adding machine learning elements on a memetic algorithm template.

Table 3.17 Reference heuristics in 10 seconds on MDPLIB 2.0

	CPLEX	C2	D2	B-VNS	ITS	TS-MA	OBMA
GKD-c (20 inst.)							
D.Best	3.83	97.35	22.27	0.00	0.00	0.02	0.00
#Best	0	0	0	19	19	1	19
GKD-d (140 inst.)							
D.Best	3.05	16.54	41.33	0.06	0.58	0.06	0.06
#Best	46	0	0	108	109	45	108
MDG-a (60 inst.)							
D.Best	-	74.51	40.59	1.18	1.19	0.01	1.14
#Best	0	0	0	22	20	58	28
MDG-b (60 inst.)							
D.Best	-	74.04	28.27	0.07	0.10	0.04	0.03
#Best	0	0	0	0	24	44	24
MDG-c (20 inst.)							
D.Best	-	99.56	76.23	0.08	0.20	0.04	0.00
#Best	0	0	0	0	0	0	20
SOM-a (50 inst.)							
D.Best	2.45	85.65	41.01	0.00	0.05	0.00	0.00
#Best	18	0	0	50	48	50	50
SOM-b (20 inst.)							
D.Best	6.40	96.45	22.73	0.00	0.00	0.00	0.00
#Best	0	0	0	20	19	20	20
SOM-c (80 inst.)							
D.Best	-	19.05	23.21	0.20	0.25	0.02	0.03
#Best	0	0	0	13	10	63	58
ALL SETS (450 inst.)							
D.Best	-	70.39	36.95	0.20	0.30	0.02	0.16
#Best	65	0	0	232	249	281	327

Martí et al. [119] compare in a second experiment the best methods above, namely D2, B-VNS, and OBMA, run with a time limit of 600 seconds per instance. Although D2 exhibits low quality results compared with the metaheuristics, we include it to show the performance of a simple constructive heuristic. Additionally, we include in this experiment the solutions obtained with CPLEX which requires on average about one hour of CPU time. Table 3.18 shows the same statistics as the previous table but, for the sake of simplicity, only on aggregate results.

Table 3.18 Best MDP methods run for 600 seconds

	CPLEX	D2	B-VNS	OBMA
D.Best	-	29.92	0.01	0.00
#Best	105	0	318	447

Results in this table show that if we apply a simple heuristic, such as D2, we obtain solutions with a relative deviation to the best known solution close to 30%.

It is clear that complex metaheuristics obtain better solutions, specially over a long period of time such as the one of 600 seconds reported in Table 3.18. However, we need also to keep in mind all the knowledge and effort required to implement such a complex method. In this experimentation, OBMA emerges as the best algorithm again, obtaining the best percentage deviation overall.

Chapter 4
Branch-and-Bound

Abstract We now turn to the discussion of how to solve the linear ordering problem
and the maximum diversity problem to (proven) optimality. In this chapter we start
with the branch-and-bound method which is a general procedure for solving combinatorial optimization problems. In the subsequent chapters this approach will be realized for the LOP in a special way leading to the so-called *branch-and-cut method*.
There are further possibilities for solving these problems exactly, e.g. by formulating them as *dynamic program* or as equivalent problems, but these approaches did
not lead to the implementation of practical algorithms and we will not elaborate on
them here.

4.1 Introduction

Combinatorial optimization deals with a special type of mathematical optimization
problem with the property that the set of feasible solutions is finite. In its most
general form such a problem is defined on a finite set $\mathscr{I}$ (set of feasible solutions)
and a function $f : \mathscr{I} \to \mathbf{R}$ has to be optimized. Since the set of feasible solutions $\mathscr{I}$
is finite, the problem could in principle be solved by enumeration. However, the
number of feasible solutions can be very large, thus prohibiting this approach in
general.

Branch-and-bound tries to deal with these many feasible solutions in a systematic
way. Basically, it is a divide-and-conquer approach that tries to solve the original
problem by splitting it into smaller problems for which upper and lower bounds
are computed and may be employed to exclude large parts of the solution set from
further consideration.

Of course, the general definition of a combinatorial optimization problem given
above is of no use unless we have a reasonable characterization of $\mathscr{I}$ and an algorithmic way of evaluating the objective function.

For many problems the objective function can be defined in a simple way and
they can be formulated as follows. (2^E denotes the power set of E.)

Definition 4.1. Let the finite set $E = \{e_1, e_2 \ldots, e_n\}$ (ground set) and $\mathscr{I} \subseteq 2^E$ (set of feasible solutions) be given. Assume that there is a function $c : E \to \mathbf{R}$ (objective function) such that the value of a feasible solution $F \in \mathscr{I}$ is given as $c(F) = \sum_{e \in F} c(e)$. The *linear combinatorial optimization problem* $(E, \mathscr{I}, c)$ consists of finding $F \in \mathscr{I}$ such that $c(F)$ is as large as possible.

The LOP fits into this scheme by setting $E = A_n$, $\mathscr{I} = \{T \subset A_n \mid T$ is an acyclic tournament$\}$, and $c((i, j)) = c_{ij}$. By complementing the function c we can also deal with minimization problems.

The crucial part of a successful branch-and-bound algorithm is the computation of upper bounds for subproblems. Here one uses the fundamental concept of relaxation.

Definition 4.2. Suppose that two combinatorial problems $(E, \mathscr{I}, f)$, $(E', \mathscr{I}', f')$ and an injective function $\varphi : E \to E'$ are given. The problem $(E', \mathscr{I}', f')$ is a *relaxation* of $(E, \mathscr{I}, f)$, if $\varphi(F) \in \mathscr{I}'$ and $f(F) = f'(\varphi(F))$, for all $F \in \mathscr{I}$.

(More generally one can define that a problem $\max\{f(x) \mid x \in T\}$ is a relaxation of the problem $\max\{c(x) \mid x \in X\}$ if $X \subseteq T$ and $f(x) \geq c(x)$, for all $x \in X$.)

Hence a solution of the relaxed problem gives an upper bound on the optimum objective function value of the problem it was derived from. The tighter the relaxation, the better this bound will be. Of course, a relaxation is only useful if it can be treated efficiently by optimization algorithms.

Branch-and-bound can be outlined as follows.

Branch-and-Bound Algorithm

(1) Initialize the list of active subproblems with the original problem.
(2) If the list of active subproblems is empty, `Stop` (the best feasible solution found so far is optimal).
(3) Otherwise, choose some subproblem from the list of active problems and "solve" it as follows:

    (3.1) Find an optimal solution for the subproblem, or
    (3.2) prove that the subproblem has no feasible solution, or
    (3.3) prove that there is no feasible solution for the subproblem that has larger objective value than the best feasible solution that is already known, or
    (3.4) split the subproblem into further subproblems and add them to the list of active problems.

(4) Goto step (2).

The splitting of problems into subproblems can be represented by the so-called *branch-and-bound tree*, the root of which represents the original problem.

It is crucial for the efficiency of a branch-and-bound algorithm that the branch-and-bound tree does not grow too large. Already, if problems are only split into two subproblems, the number of subproblems grows exponentially fast. Therefore subproblems have to be solved if possible by alternatives (3.1), (3.2) or (3.3). Alternative (3.1) rarely occurs, and relaxations are important for (3.2) and (3.3). With respect to (3.2), if a relaxation of the subproblem is already infeasible, then the subproblem itself is also infeasible. To be able to fathom the subproblem using (3.3), good lower and upper bounds must be available. Lower bounds are obtained by finding feasible solutions. These are computed either by solving some subproblem to optimality or more often by determining good feasible solutions using one of the many heuristics we have discussed in the previous chapters. Upper bounds can be computed by using relaxations where in principle any type of relaxation can be employed. It is clear that the stronger the bounds obtained by a relaxation are, the better the performance of the algorithm will be. Without a suitable relaxation branch-and-bound tends to completely enumerate the set of feasible solutions and thus becomes unsuitable for practical computations.

Besides *bounding*, the second component of this approach is *branching* which denotes the splitting of the current subproblem into a collection of new subproblems whose union of feasible solutions contains all feasible solutions of the current subproblem. For 0/1-problems, the simplest branching rule consists of defining two new subproblems: in the first subproblem some chosen variable is required to have the value 1 in every feasible solution and in the second one to have the value 0. Other branching strategies are possible. There are also several heuristics for choosing the next subproblem to be considered.

4.2 Branch-and-Bound with Partial Orderings for the LOP

One of the earliest branch-and-bound algorithms for the LOP was proposed by de Cani [46] in 1972. He encountered the linear ordering problem when studying procedures to obtain a ranking of n objects on the basis of a number of pairwise comparisons. The algorithm successively constructs partial rankings of more and more objects and tries to prove by upper bounds that some partial rankings need not be extended for finding optimum solutions.

An upper bound for the optimum solution value of a LOP is clearly given by

$$z_0 = \sum_{i=1}^{n-1} \sum_{j=i+1}^{n} \max\{c_{ij}, c_{ji}\}.$$

If we require that object i is to be ranked before object j then an upper bound is given by

$$z(i, j) = z_0 - \max\{c_{ij}, c_{ji}\} + c_{ij}.$$

Partial rankings are built up, and each branching operation in the tree corresponds to inserting a further object at some position in the partial ranking. For each partial ranking an upper bound for the best possible extension to a complete ranking can be calculated.

Suppose that an ordering of k objects (w.l.o.g. objects $1,\ldots,k$), say $\langle 1,2,\ldots,k\rangle$, is given. If we insert object $k+1$ at position l, $1 \leq l \leq k+1$, of this ordering then, as an upper bound on the objective function value for the linear orderings containing the respective partial ordering of $k+1$ objects, we get

$$z(1,\ldots,l-1,k+1,l,l+1,\ldots,k) =$$

$$z(1,\ldots,k) + \sum_{r=1}^{l-1} c_{r,k+1} + \sum_{r=l}^{k} c_{k+1,r} - \sum_{r=1}^{k} \max\{c_{k+1,r}, c_{r,k+1}\}.$$

We start the branch-and-bound algorithm by arbitrarily choosing two of the objects, say i and j, and calculate $z(i,j)$ and $z(j,i)$. We generate two nodes, the first one corresponding to the partial ordering ranking i before j, and the second one to ranking j before i. Then we proceed at the node with the larger upper bound value. Suppose that we are at a node of level k, $1 < k < n$, of the tree (assuming that the root node is on level 1). Then k objects are already partially ordered. From the remaining $n-k$ objects we select one (according to some rule). The $k+1$ upper bounds obtained by inserting the new object at each possible position are calculated, and we proceed in that branch of the tree corresponding to the largest of these values. At level n a complete ranking of the objects is found.

The upper bounds are exploited in the usual way for backtracking and excluding parts of the tree from further consideration.

Figure 4.1 visualizes the development of the branch-and-bound tree with this approach. Nodes are labeled with partial ordering relations. Note that it is not a binary tree, the number of branches increases with the level.

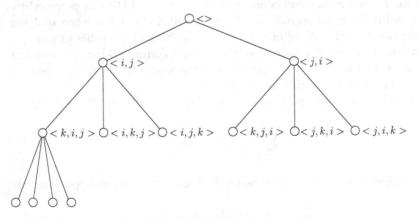

Fig. 4.1 Construction of the branch-and-bound tree

4.3 Lexicographic Search

Another one of the early methods for the solution of linear ordering problems is the *lexicographic search algorithm* proposed in 1968 by Korte and Oberhofer [101, 102]. It is actually not a branch-and-bound algorithm and can better be characterized as an *enumeration scheme*. Korte and Oberhofer were interested not only in an optimum triangulation of an input-output matrix, but also in the number of optima and the number of so-called relatively optimum solutions.

Definition 4.3. A matrix $C = (c_{ij})$ satisfies *Helmstädter's conditions* if

(i) $\sum_{l=i}^{k} c_{il} \geq \sum_{l=i}^{k} c_{li}$, for all $i \leq k$, and

(ii) $\sum_{l=k}^{i} c_{li} \geq \sum_{l=k}^{i} c_{il}$, for all $k \leq i$.

Such matrices are also called *relatively optimum*.

These conditions were given by Helmstädter [82]. Of course, every optimally triangulated matrix satisfies *Helmstädter's conditions*. If one of the conditions is violated, then a simple reordering can improve the objective function. If, for example, condition (i) is violated for some i and k, then it would be profitable to change the subsequence $\langle i, i+1, \ldots, k \rangle$ to $\langle i+1, \ldots, k, i \rangle$.

The lexicographic search algorithm enumerates all permutations of the n objects by fixing at level k of the enumeration tree the k-th position of the permutations. More precisely, if a node at level k is generated, then the first k positions $\sigma(1), \ldots, \sigma(k)$ are fixed. Based on this fixing, several of Helmstädter's conditions can be checked. If one is violated then no relatively optimum solution has $\sigma(1), \ldots, \sigma(k)$ in the first k positions. The node can be ignored, and a backtracking operation is performed.

Figure 4.2 shows part of the enumeration tree for 4 objects. Nodes at level k show the fixings of the first k positions. (Here the root node is at level 0.)

Since there is no bounding according to objective function values eventually all relatively optimum solutions are enumerated, and the best ones among them are optimum. Korte and Oberhofer applied their algorithm to get information about the distribution of relatively and absolutely optimum solutions for real-world and random problems.

4.4 Extension of Lexicographic Search to Branch-and-Bound

Possiblities for the derivation of bounds are already mentioned in [102], although not implemented. Recall that we may w.l.o.g. assume that all matrix entries c_{ij} are nonnegative.

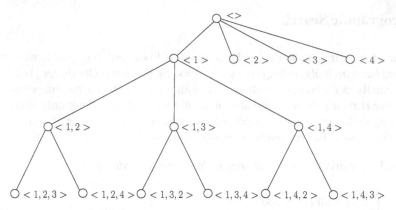

Fig. 4.2 Lexicographic search tree

Suppose that we have fixed the first k positions $\sigma(1), \ldots, \sigma(k)$ of a permutation. Let $I = \{\sigma(1), \ldots, \sigma(k)\}$ and $J = \{1, \ldots, n\} \setminus I$. If σ is extended to some complete permutation of $\{1, \ldots, n\}$ we have

$$\sum_{i=1}^{n-1} \sum_{j=i+1}^{n} c_{\sigma(i)\sigma(j)} = \sum_{i=1}^{k-1} \sum_{j=i+1}^{k} c_{\sigma(i)\sigma(j)} + \sum_{i \in I} \sum_{j \in J} c_{ij} + \sum_{i=k+1}^{n-1} \sum_{j=i+1}^{n} c_{\sigma(i)\sigma(j)} .$$

The first two terms on the right hand side of this equation do not depend on the permutation of J. An upper bound for the third term (which amounts to determining an upper bound for the triangulation problem with respect to the ground set J) yields an upper bound for all possible extensions.

Kaas [94] developed a lexicographic branch-and-bound scheme based on the algorithm of [102]. He also applies Helmstädter's conditions to rule out solutions which are not relatively optimum, but in addition uses this equation to compute upper bounds (heuristics are used for bounding the third term from above).

Note that the sum of the subdiagonal entries of the matrix obtained by permuting the rows and columns according to σ is calculated by

$$\sum_{i=1}^{n-1} \sum_{j=i+1}^{n} c_{\sigma(j)\sigma(i)} = \sum_{i=1}^{k-1} \sum_{j=i+1}^{k} c_{\sigma(j)\sigma(i)} + \sum_{i \in I} \sum_{j \in J} c_{ji} + \sum_{i=k+1}^{n-1} \sum_{j=i+1}^{n} c_{\sigma(j)\sigma(i)} .$$

The sum of the first two right-hand side terms gives a lower bound on the sum of the subdiagonal entries of any extension and can thus be used to compute an upper bound on the sum of the superdiagonal entries.

There are more authors who have formulated a branch-and-bound method for this problem (e.g. [81, 107, 140]) based on the same ideas as outlined above.

4.5 Branch-and-Bound with Lagrangean Relaxation

This is a more recent branch-and-bound approach of Charon and Hudry [34] where Lagrangean relaxation techniques are used for bound computations.

Define 0/1 variables x_{ij}, $1 \leq i, j \leq n$, $i \neq j$, where $x_{ij} = 1$ if i is ranked before j, and $x_{ij} = 0$ otherwise. Then the LOP can be formulated as the linear 0/1 program

$$\max \sum_{(i,j) \in A_n} c_{ij} x_{ij}$$

$$x_{ij} + x_{jk} + x_{ki} \leq 2, \text{ for all distinct nodes } i, j, k \in V_n,$$
$$x_{ij} + x_{ji} = 1, \text{ for } 1 \leq i < j \leq n,$$
$$x_{ij} \in \{0, 1\}, \text{ for } 1 \leq i, j \leq n, i \neq j.$$

This is the canonical IP formulation of the LOP and we will discuss it in more detail in the next chapters.

Lagrangean relaxation removes constraints from the original problem and penalizes their violation in a modified objective function.

Let T be the set of all triples (r, s, t) of three distinct nodes where $r < s$ and $r < t$. The following Lagrangean relaxation is used in [34] for a given vector $\mu \geq 0$ of Lagrangean multipliers.

$$L(\mu) = \max \sum_{(i,j) \in A_n} c_{ij} x_{ij} + \sum_{(r,s,t) \in T} (2 - x_{rs} - x_{st} - x_{tr}) \mu_{rst}$$

$$x_{ij} + x_{ji} = 1, \text{ for } 1 \leq i < j \leq n,$$
$$x_{ij} \in \{0, 1\}, \text{ for } 1 \leq i, j \leq n, i \neq j.$$

This problem can be solved trivially (more details below) and obviously gives an upper bound on the optimum objective function value of the LOP. Note that this relaxation does not exactly meet the requirements of Definition 4.2. Here we have $f(F) \leq f'(\varphi(F))$, but the upper bound property holds as well.

The best such bound can be found by solving the *Lagrangean dual problem*

$$\min_{\mu \geq 0} L(\mu).$$

To this end one uses so-called *subgradient* or *bundle methods* which are able to find good approximations of the best bound. In theory, they could compute this bound exactly, but then a very slow convergence of the step sizes of the subgradient algorithm to 0 is required. In practice, step sizes are decreased faster and the optimum bound is therefore not met exactly.

With the notation $(i, j) \in (r, s, t)$ if (i, j) is one of the edges (r, s), (s, t) or (t, r) the objective function for computing $L(\mu)$ can be rewritten as

$$\max \sum_{(i,j)\in A_n} c_{ij}x_{ij} + \sum_{(i,j,k)\in T} (2 - x_{ij} - x_{jk} - x_{ki})\mu_{ijk}$$

$$= \max \sum_{(i,j)\in A_n} \left(c_{ij} - \sum_{(r,s,t)\in T,(i,j)\in(r,s,t)} \mu_{rst}\right)x_{ij} + 2\sum_{(r,s,t)\in T} \mu_{rst}$$

$$= \max \sum_{(i,j)\in A_n} d_{ij}x_{ij} - C,$$

where d_{ij} is set according to the second-to-last line and C is constant. In this form it is clear that finding the optimum in 0/1 variables with only $x_{ij} + x_{ji} = 1$ as constraints is trivial.

We now describe a prototypical realization of a subgradient approach for approximating $\min_\mu L(\mu)$.

LOPSubgradient(D_n, c)

(1) Let τ be an initial step size and $0.9 \leq \alpha < 1$ a decrement factor.

(2) Set $t_1 = \tau$, $\mu_{rst}^0 = 0$ for every $(r,s,t) \in T$ and $k = 1$.

(3) While $t_k > \varepsilon$:

    (3.1) Compute $L(\mu)$ by setting $x_{ij} = 1$, if $d_{ij} > d_{ji}$, and $x_{ij} = 0$, otherwise, for $i < j$, and set $x_{ji} = 1 - x_{ij}$.

    (3.2) Define d^k by setting $d_{rst}^k = 2 - x_{rs} - x_{st} - x_{tr}$, for $(r,s,t) \in T$.

    (3.3) Set $\mu_{rst}^{k+1} = \mu_{rst}^k + t_k \times d_{rst}^k$, for every $(r,s,t) \in T$. If $\mu_{rst}^{k+1} < 0$, then set $\mu_{rst}^{k+1} = 0$.

    (3.4) Set $t_{k+1} = \alpha t_k$ and increment k.

(4) Return the best bound found.

We do not want to introduce the background of nondifferentiable optimization here. We just note that d^k is a so-called *subgradient* and that with α very close to 1 and ε small, $\min_\mu L(\mu)$ is usually very well approximated at the expense of, however, considerable running time. The decrease of the bounds is not monotone, and therefore in (4) the best bound found is returned. Bundle methods extend this principle and compute combinations of subgradients of several iterations for the direction of the next step. In practice, they are much more powerful.

We illustrate this (simple) subgradient algorithm with two applications for finding an upper bound for the LOLIB instance be75np. In the first run (Fig. 4.3) α is set to 0.999999, while in the second run (Fig. 4.4) it is set to 0.99. The figures show the development of the upper bounds for the first 100 iterations. The first run terminates after 2 hours giving the upper bound 790989 (which as will be seen in the next chapter is the minimum of $L(\mu)$), the second run terminates after only 5 seconds giving the upper bound 799467. The decrease of the upper bound is fast in the first iterations. After 100 iterations we have bound 800077 in the first run and 801963 in the second run, but to reach bounds near the optimum substantial running time has to be invested.

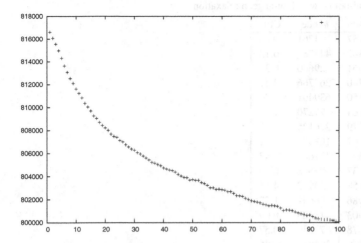

Fig. 4.3 Simple subgradient method with slow convergence

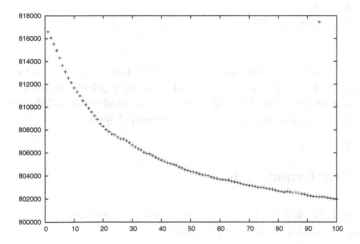

Fig. 4.4 Simple subgradient method with fast convergence

This is just a simple example for the general behaviour of subgradient methods. In [34] a variant that finds a good compromise between fairly fast convergence and good quality of bounds is developed. With this branch-and-bound method, instances of moderate sizes can be solved successfully. As the heuristic for finding good linear orderings, the meta-heuristic *noising method* is used which was described in Chap. 3.

The algorithm is publicly available [34] and we have applied it (with standard parameter settings) to a set of random benchmark instances with $n = 40$. Table 4.1

Table 4.1 Branch-and-bound with Lagrangean relaxation

Problem	Optimum	#nodes	CPU
p40-01	29457	4261	0:06
p40-02	27482	444285	6:13
p40-03	28061	29660	0:31
p40-04	28740	267766	3:51
p40-05	27450	639401	9:26
p40-06	29164	79270	1:09
p40-07	28379	226125	3:27
p40-08	28267	103477	1:31
p40-09	30578	431607	5:47
p40-10	31737	86698	0:59
p40-11	30658	333902	4:16
p40-12	30986	96651	1:27
p40-13	33903	13423	0:09
p40-14	34078	73839	0:57
p40-15	34659	396839	4:45
p40-16	36044	9295	0:07
p40-17	38201	3040	0:04
p40-18	37562	9669	0:07
p40-19	38956	9857	0:07
p40-20	39658	13067	0:09

shows the number of branch-and-bound nodes and the CPU time on a standard PC. Large variations can be observed which are depending on the gap between the Lagrangean bound and optimum value. We will compare these results with a different bounding approach based on linear programming in the next chapter.

4.6 Improved MDP formulations

As shown in Chapter 1, Section 1.2, the maximum diversity problem can be formulated as a linear integer problem with binary variables:

$$(F1)\max \sum_{i<j} d_{ij}x_i x_j$$

$$\sum_{i=1}^{n} x_i = m$$

$$x_i \in \{0,1\}, \; i = 1,\ldots,n.$$

To avoid the non-linearity due to the product of two variables, Kuby [54] formulated the MDP in 1988 as:

$$\max \sum_{i<j} z_{ij} d_{ij}$$

$$\sum_{i=1}^{n} x_i = m$$

$$z_{ij} \leq x_i, \quad i,j = 1,\dots,n : j > i$$
$$z_{ij} \leq x_j, \quad i,j = 1,\dots,n : j > i$$
$$z_{ij}, x_i \in \{0,1\}, \quad i,j = 1,\dots,n.$$

The z_{ij} binary variables correspond to the product of the $x_i x_j$. A similar approach was proposed a few years later by Kuo et al. [104] that apply a standard transformation to the quadratic formulation to replace the product $x_i x_j$ with a new variable y_{ij}, obtaining the mixed integer program below, in which some additional constraints have been included.

$$(F2)\max \sum_{i<j} d_{ij} y_{ij}$$

$$\sum_{i=1}^{n} x_i = m$$

$$x_i + x_j - y_{ij} \leq 1, \quad i,j = 1,\dots,n : i < j$$
$$-x_i + y_{ij} \leq 0, \quad i,j = 1,\dots,n : i < j$$
$$-x_j + y_{ij} \leq 0, \quad i,j = 1,\dots,n : i < j$$
$$y_{ij} \geq 0, \quad i,j = 1,\dots,n : i < j$$
$$x_i \in \{0,1\}, \quad i = 1,\dots,n.$$

An alternative formulation is also proposed in [104] by decomposing the objective function z in formulation (F1) into the sum of w-values as:

$$z = \sum_{i=1}^{n-1} \left(x_i \sum_{j=i+1}^{n} d_{ij} x_j \right) = \sum_{i=1}^{n-1} w_i.$$

The authors obtained formulation (F3) with only two constraints,

$$-D_i x_i + w_i \leq 0 \quad , \quad - \sum_{j=i+1}^{n} d_{ij} x_j + E_i (1 - x_i) + w_i \leq 0$$

where the values of D_i and E_i are set as:

$$D_i = \sum_{j=i+1}^{n} max(0, d_{ij}) \text{ and } E_i = \sum_{j=i+1}^{n} min(0, d_{ij}).$$

apart from the standard constraint to set the number of elements equal to m.

To see the equivalence with the initial formulation, we can see that when $x_i = 0$, then the first new constraint implies that $w_i \leq 0$, and the second new constraint results in

$$-\sum_{j=i+1}^{n} d_{ij}x_j + E_i + w_i \le 0$$

which implies that

$$w_i \le \sum_{j=i+1}^{n} d_{ij}x_j - E_i.$$

Since by construction

$$E_i \le \sum_{j=i+1}^{n} d_{ij}x_j,$$

then the maximum value that w_i can take is 0. Similarly, we can see that if $x_i = 1$, then

$$w_i = \sum_{j=i+1}^{n} d_{ij}x_j,$$

thus making (F1) and (F3) equivalent in terms of the objective function.

$$(F3) \max \sum_{i=1}^{n-1} w_i$$

$$\sum_{i=1}^{n} x_i = m$$

$$-D_i x_i + w_i \le 0, \quad i = 1,\ldots,n-1$$

$$-\sum_{j=i+1}^{n} d_{ij}x_j + E_i(1-x_i) + w_i \le 0, \quad i = 1,\ldots,n-1$$

$$x_i \in \{0,1\}, \quad i = 1,\ldots,n.$$

Martí et al. [115] compared the three linear integer formulations, (F1), (F2) and (F3), described above on a Pentium 4 computer at 3 GHz with 3 GB of RAM. The authors limited the execution of the CPLEX 8.0 solver to 1 hour (3,600 seconds) of computer time. For each combination of the n and m values in the 75 Euclidean instances tested, Table 4.2 shows the average gap, Gap, and the CPU time in seconds, for each formulation. The average gap is computed as the upper bound minus the best solution found (both returned by CPLEX when the time limit is reached), divided by the upper bound and multiplied by 100.

Table 4.2 shows that formulation F3 produces better results than (F1) and (F2) with respect to CPU time. Specifically, the application of (F1) and (F2) leads to average gaps of 0.7% and 0.8% and the computer solution times are 591.6 and 711.7 seconds on average respectively, while the application of (F3) leads to an average gap of 0.0% and the computer solution time is 96.9 seconds on average. In addition, (F1), (F2), and (F3) obtain 65, 66, and 75 optimal solutions respectively (out of 75 instances).

Table 4.2 Comparison of MDP formulations

n	m	F1 Gap	F1 CPU	F2 Gap	F2 CPU	F3 Gap	F3 CPU
10	2	0.0	0.0	0.0	0.1	0.0	0.0
10	3	0.0	0.0	0.0	0.1	0.0	0.0
10	4	0.0	0.0	0.0	0.1	0.0	0.0
10	6	0.0	0.0	0.0	0.1	0.0	0.0
10	7	0.0	0.0	0.0	0.1	0.0	0.0
15	3	0.0	0.0	0.0	0.6	0.0	0.1
15	4	0.0	0.1	0.0	0.8	0.0	0.1
15	6	0.0	0.4	0.0	1.3	0.0	0.2
15	9	0.0	0.4	0.0	1.1	0.0	0.2
15	12	0.0	0.0	0.0	0.6	0.0	0.0
30	6	0.0	59.4	0.0	514.7	0.0	35.7
30	9	0.0	1597.0	4.7	3564.7	0.0	292.9
30	12	7.6	3600.0	6.8	3501.9	0.0	929.3
30	18	3.5	3600.0	1.0	3055.9	0.0	193.2
30	24	0.0	16.8	0.0	33.7	0.0	1.9
Average		0.7	591.6	0.8	711.7	0.0	96.9

More recently, Parreño et al. [138] tested formulation (F3) in the entire MDPLIB benchmark set of 675 instances to evaluate the ability of this formulation implemented in CPLEX 12.8 to match optimal solutions within a time limit of 1 hour of CPU. The authors excluded from this analysis the original sets MDG in the MDPLIB because they are too large to be solved with CPLEX, and included two new sets MDG-a2 and MDG-b2 with 20 instances, each one of size $n = 100$. Table 4.3 shows the results of their experiment, in which for each instance set it reports the number of instances, Num., their size, n, the number of optimal solutions found, Opts., and the average gap, Gap, computed as in the previous experiment shown above.

Table 4.3 Results with F3 by instance sets

Set	Num.	n	Opts.	Gap
GKD-a	75	[10, 30]	75	0.0
GKD-b	50	[25, 150]	15	79.3
GKD-c	20	500	0	584.5
GKD-d	300	[25, 500]	100	136.0
MDG-a2	20	100	0	114.9
MDG-b2	20	100	0	129.9
SOM-a	50	[25,150]	15	76.2
SOM-b	20	[100, 500]	0	158.9
Total	675	[10, 500]	205	213.3

Table 4.3 clearly shows that some types of instances are more difficult to solve than others, considering that the average gap obtained is significantly larger. That

is the case for example of GKD-c with an average Gap of 584.5, which is larger than the gap obtained in the set GKD-d of 136.0. Regarding problem size, most of the sets contain instances of medium size ($n = 100$), with the exception of the GKD-a that only contains very small instances ($n \leq 30$), and GKD-c with only large instances ($n = 500$). As expected, the method is able to solve to optimality all the instances in GKD-a, and none of them in GKD-c. Note that the instances in sets MDG-a2 and MDG-b2 are all of them of size $n = 100$, and CPLEX is not able to identify any optimal solution. We may conclude that there is still plenty of room for improvement in terms of strengthen this formulation to be able to solve, at least, medium size problems.

4.7 Branch-and-Bound with Partial Selections for the MDP

In this section we summarize the work in Martí et al. [115], who proposed a branch and bound for the MDP based on partial selections (solutions), and which compares favorably with the performance of CPLEX with the formulation (F3) shown in the previos section.

Given a graph $G_n = (V_n, E_n)$ on n nodes, with distance d_{ij} on edge (i, j), a solution of the MDP is a set $M \subset V_n$ of m elements, and its value is the sum of distances between the pairs of nodes in M. A partial solution Sel is a set of k nodes in V_n with $k < m$. The set $Sel \subset V$ can be viewed as an incomplete solution. We can obtain a complete solution to the MDP by adding $m - k$ elements from $V \setminus Sel$ to it. Let SEL be the set of all solutions obtained by adding elements to Sel (i.e. the set of solutions in which all the elements in Sel are selected).

Consider a partial solution $Sel = \{s_1, s_2, \ldots, s_k\}$ and a complete solution $x = \{s_1, s_2, \ldots, s_k, u_1, u_2, \ldots, u_{m-k}\}$ in SEL. We denote the set of vertices in x not present (unselected) in Sel as $U(x) = \{u_1, u_2, \ldots, u_{m-k}\}$. The objective function value z associated with x can be broken down into $z = z_1 + z_2 + z_3$, where z_1 is the sum of the distances (edge weights) between the pairs of selected vertices (vertices in Sel), z_2 is the sum of the edge weights with one extreme in Sel and the other in $U(x)$, and z_3 is the sum of edge weights with both extremes in $U(x)$. More specifically,

$$z_1 = \sum_{i=1}^{k-1} \sum_{j=i+1}^{k} d_{s_i s_j} \quad , \quad z_2 = \sum_{i=1}^{k-1} \sum_{j=1}^{m-k} d_{s_i u_j} \quad , \quad z_3 = \sum_{i=1}^{m-k-1} \sum_{j=i+1}^{m-k} d_{u_i u_j}.$$

We now compute an upper bound of the objective function value of the solutions in SEL. Note that z_1 is an invariant in SEL and therefore, we only need to obtain an upper bound on $z_2 + z_3$.

Given a partial solution $Sel = \{s_1, s_2, \ldots, s_k\}$, and a vertex $v \in V \setminus Sel$, we define $z_{Sel}(v)$ as

$$z_{Sel}(v) = \sum_{i=1}^{k} d_{s_i v}.$$

We may consider $z_{Sel}(v)$ as the potential contribution of v with respect to the selected vertices if we add it to the partial solution Sel.

Let $d^1(v), d^2(v), \ldots, d^{n-k-1}(v)$ be the distances from v to the other $n - k - 1$ nodes in $V \setminus Sel$. We define $z_U(v)$ as

$$z_U(v) = 1/2 \sum_{i=1}^{m-k-1} d^i(v).$$

We may consider $z_U(v)$ as an upper bound on the potential contribution of v with respect to the unselected vertices, if we add it to the partial solution Sel under construction. Therefore $z(v) = z_{Sel}(v) + z_U(v)$ provides an upper bound of the contribution of v if added to the partial solution.

Let $z^1, z^2, \ldots, z^{n-k}$ be the values of $z(v)$ in descending order for all v nodes in $V \setminus Sel$. If we add up the $m - k$ largest ones, we obtain an upper bound on the value of $z_2 + z_3$. Then, an upper bound of the value of all the solutions in SEL is given by the expression:

$$UB = z_1 + \sum_{i=1}^{m-k} z^i.$$

Based on this upper bound, Martí et al. [115] proposed a branch and bound for the MDP. The algorithm iterates over a search tree that provides a generation and partition of the set of solutions in which each node represents a partial solution (except the leaves or final nodes that represent complete solutions).

The initial node branches into n nodes (labeled from 1 to n, where node i represents the partial solution $Sel = \{i\}$). Each of these n nodes in the first level branches into $n - 1$ nodes (which will be referred to as nodes in level 2). For instance, node 2 in the first level ($Sel = \{2\}$) has $n - 1$ successors in level 2 (labeled as 1, 3, 4,..., n). So, node 3 in the second level, the successor of node 2 in the first level, represents the partial solution $Sel = \{2, 3\}$. Therefore, at each level in the search tree, the algorithm extends the current partial solution by adding one vertex.

The basic search tree described above contains repetitions of the same solutions. Therefore, this tree does not represent a partition of the set of complete solutions to the MDP. We can improve it, making the search more efficient, in the following way. The initial node branches into $n-m+1$ nodes (labeled from 1 to $n-m+1$, where node i represents the partial solution $Sel = \{i\}$). Each of these $n-m+1$ nodes in the first level branches into a number of nodes in level 2 that depends on the node label. In general, node i in level k branches into $n - (m - k)-i+1$ nodes, beginning with node $i+1$ and ending with node $n - (m - k) + 1$.

Figure 4.5 shows this search tree with no repetitions of an example with $n = 5$ and $m = 3$. Node 1 in level 1 branches into nodes 2, 3 and 4; node 2 in level 1 branches into nodes 3 and 4, and node 3 in level 1 branches into node 4. Node 2 in level 2 branches into nodes 3, 4 and 5; nodes 3 branch into nodes 4 and 5, and nodes 4 branch into node 5. The nodes in level 3, leaf nodes, represent complete solutions that are shown in the right hand column. Comparing the search tree of this figure with the basic design described above, we can see that the size has been substantially reduced by avoiding the repetitions.

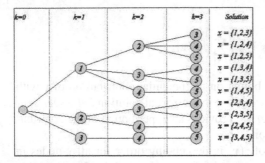

Fig. 4.5 Improved search tree

The authors further reduce the size of the search tree by proving a properties based on the potential contributions that optimal solutions have to verify. We summarize now in Table the empirical comparison between the branch and bound resulting from exploring this search tree based on partial selections, BB, and the one implemented in CPLEX 8.0 that is based on the binary mathematical programming formulation (F3).

Table 4.4 reports the results obtained with both methods, BB and CPLEX(F3), over the two sets of medium size instances, GKD-b and SOM-b. Each row displays the results for a group of five instances (with the same n and m values). As in the previous tables, we report the average gap, the CPU time in seconds, and the number of optima that each method is able to match. Note that some running times are a little longer than the time limit, especially in the CPU column under F3, since we check the running time when a complete iteration of the method is performed.

Table 4.4 shows that the CPLEX with the F3 formulation is able to solve the medium-sized instances (up to $n = 50$ on both set of instances) within 1 hour of CPU time. On the other hand, the combinatorial branch and bound algorithm BB outperforms F3 since it is able to optimally solve all the medium-size instances and some of the large instances ($n = 100$ and $m = 10$). Moreover, considering the 100 instances reported in Table 4.4, BB presents an average gap of 8.4% achieved in 1502.7 seconds, which compares favourably with the 39.5% obtained with F3 in 2558.5 seconds on average.

Table 4.4 Comparison of branch and bound methods

Set	n	m	BB Gap	BB CPU	BB Opts.	F3 Gap	F3 CPU	F3 Opts.
	25	2	0.0	0.0	5	0.0	0.2	5
	25	7	0.0	0.0	5	0.0	15.0	5
	50	5	0.0	0.0	5	0.0	462.2	5
	50	15	0.0	0.4	5	27.7	3609.7	0
GKD-b	100	10	0.0	4.4	5	71.5	3609.0	0
	100	30	8.6	3576.2	1	41.9	3603.8	0
	125	12	0.0	297.9	5	75.2	3600.0	0
	125	37	13.7	3600.0	0	45.3	3600.0	0
	150	15	5.4	1834.4	3	77.7	3600.0	0
	150	45	10.9	3600.1	0	52.7	3600.0	0
	25	2	0.0	0.0	5	0.0	0.1	5
	25	7	0.0	0.0	5	0.0	4.5	5
	50	5	0.0	0.0	5	0.0	265.7	5
	50	15	0.0	22.8	5	25.8	3600.0	0
SOM-b	100	10	0.0	38.3	5	71.1	3600.0	0
	100	30	31.7	3600.0	0	46.4	3600.0	0
	125	12	0.0	2678.9	5	76.1	3600.0	0
	125	37	34.6	3600.0	0	50.1	3600.0	0
	150	15	26.7	3600.1	0	78.1	3600.0	0
	150	45	35.6	3600.1	0	50.8	3600.0	0
Average			8.4	1502.7	3.2	39.5	2558.5	1.5

Chapter 5
Branch-and-Cut for the LOP

Abstract This chapter focuses on the approach for solving the LOP to optimality which can currently be seen as the most successful one. It is a branch-and-bound algorithm, where the upper bounds are computed using linear programming relaxations.

5.1 Integer Programming

Linear programming is concerned with the problem of maximizing a linear objective function subject to finitely many linear constraints. Given a matrix $A \in \mathbf{R}^{(m,n)}$ and vectors $b \in \mathbf{R}^m$, $c \in \mathbf{R}^n$, the task is to find a vector $x^\star \in \mathbf{R}^n$ with

$$c^T x^\star = \max\{c^T x \mid Ax \le b\}.$$

Such a problem is called a *linear programming problem* or *linear program* (*LP*). Note that a minimization problem can be transformed to a maximization problem by complementing the objective function. Equations can be expressed as pairs of inequalities, and bounds on variables can be included into $Ax \le b$. The set of feasible solutions $\{x \mid Ax \le b\}$ is a polyhedron and if the problem has a finite optimum solution then it has an optimum vertex solution. We do not go into more details on polyhedra here. They will be a central topic of Chap. 6.

Very effective algorithms for solving linear programming problems have been developed and very large instances can be treated in reasonable time. Important methods are the primal and dual simplex algorithms (as vertex following methods) and the barrier method (as an interior-point method).

For combinatorial optimization, linear programming models are not sufficient. Usually, some or all of the variables have to take integer values. If some variables have to be integral, we speak about a *mixed-integer linear programming problem*. If all variables are required to be integer, then we have an *integer linear program*; if all variables have to take values 0 or 1, the problem is called *linear 0/1 program-*

© Springer-Verlag GmbH Germany, part of Springer Nature 2022
R. Martí and G. Reinelt, *Exact and Heuristic Methods in Combinatorial Optimization*, Applied Mathematical Sciences 175, https://doi.org/10.1007/978-3-662-64877-3_5

ming problem (0/1-IP). This type of problem is an important one in combinatorial optimization. There has been significant progress in the development of algorithms for solving mixed-integer programming problems. But still, problems with several hundreds of variables and constraints can be difficult depending on their specific structure.

Usually, a linear combinatorial optimization problem $(E, \mathscr{I}, c)$ can easily be turned into a linear 0/1 programming problem. First, we associate with every set $F \subseteq E$ its *characteristic vector* χ^F by setting

$$\chi^F_e = \begin{cases} 1, & \text{if } e \in F, \\ \\ 0, & \text{otherwise.} \end{cases}$$

Then equations and inequalities have to be found which are satisfied by all characteristic vectors corresponding to the feasible sets and are violated by all 0/1 vectors for sets $F \subseteq E$ with $F \notin \mathscr{I}$. In most cases, the combinatorial properties discriminating feasible solutions can be expressed in a straightforward way by linear constraints.

The LOP can be formulated as 0/1-IP as follows. We use 0/1 variables x_{ij}, for $(i, j) \in A_n$, stating whether arc (i, j) is present in the tournament or not. The basic observation is that in a tournament exactly one of the arcs (i, j) and (j, i) is present for every pair of nodes i and j, and that a tournament is acyclic if and only if it does not contain any dicycle of length 3.

The respective 0/1-IP is

$$\max \sum_{(i,j) \in A_n} c_{ij} x_{ij}$$

$$x_{ij} + x_{ji} = 1, \ i, j \in V_n, i < j,$$
$$x_{ij} + x_{jk} + x_{ki} \leq 2, \ i, j, k \in V_n, i < j, i < k, j \neq k,$$
$$x_{ij} \in \{0, 1\}, \ i, j \in V_n.$$

This 0/1-IP can be considered as the canonical IP formulation of the problem. Note that the $\binom{n}{2}$ equations could actually be simply removed by substituting every variable x_{ij}, $j > i$, by $1 - x_{ji}$. More problematical is the large number $2\binom{n}{3}$ of 3-dicycle constraints.

Obviously, if we replace the constraints "$x_{ij} \in \{0, 1\}$" by "$0 \leq x_{ij} \leq 1$" then we obtain a linear programming problem, the *canonical LP relaxation*. The optimum objective function value of this relaxation gives an upper bound on the solution value of the LOP.

If, by chance, the optimum solution is integral then the characteristic vector of an optimum acyclic tournament is found. If not, then together with heuristics providing good feasible solutions, we can immediately design a branch-and-bound algorithm for solving the LOP to optimality.

In this branch-and-bound approach problems are split into subproblems by fixing one selected variable either to 1 or to 0. During its execution the algorithm keeps a list $\mathcal{K}$ of active problems and the value L of the best feasible solution $\bar{x}$ found so far.

LOP Branch-and-Bound

(0) Run a heuristic to provide a feasible ordering $\bar{x}$ with value L.

(1) Let problem P_0 be the canonical IP formulation. Set $\mathcal{K} = \{P_0\}$ and $k = 0$.

(2) If $\mathcal{K} = \emptyset$, Stop (The ordering $\bar{x}$ is optimal with value L).

(3) Select a problem $P_j \in \mathcal{K}$.

(4) Solve the linear programming relaxation LP_j of P_j. If LP_j is infeasible then set $c^* = -\infty$, otherwise let x^* be its optimum solution with value c^*. Distinguish the following three cases:

   (4.1) If $c^* \leq L$, then remove P_j from $\mathcal{K}$.

   (4.2) If $c^* > L$ and x_i^* is integral, then set $L = c^*$, $\bar{x} = x^*$ and remove P_j from $\mathcal{K}$.

   (4.3) If $c^* > L$ and x_i^* is not integral, then select an arc $(i, j) \in A_n$ with fractional value x_{ij}^*. Remove P_j from $\mathcal{K}$ and add the new problems P_{k+1} and P_{k+2} to $\mathcal{K}$, where

          $P_{k+1} = P_j$ with additional constraint $x_{ij} = 1$,
          $P_{k+2} = P_j$ with additional constraint $x_{ij} = 0$.
          Set $k = k+2$.

(5) Goto (2).

Clearly, this algorithm solves the LOP in finite time since in the worst case all possible feasible solutions would be enumerated. In principle, this approach actually splits a problem by putting bounds on a single variable and can also be used for general linear 0/1 or mixed integer programs. It was originally formulated for the latter case by Dakin [43].

The above enumeration scheme essentially fixes $i \prec j$ in one branch and $j \prec i$ in the other one. Of course, other schemes could also be used in this framework.

5.2 Cutting Plane Algorithms

If the LOP is formulated as 0/1-IP as above, instances of moderate size, say up to $n = 30$, can be solved with a commercial MIP solver in several hours. The solution time is highly problem dependent, in particular depending on the strength of the LP relaxation. For easy instances, the relaxation can sometimes even already provide an optimum solution. For difficult instances, however, the LP bound can be more than 4% off the optimum value and a large number of branch-and-bound nodes is

generated. It is the main purpose of this and the following chapter to exhibit ways
of overcoming these difficulties.

A first observation is that the number of dicycle constraints $2\binom{n}{3}$ is fairly large
(already 323,400 for $n = 100$). It constitutes a major drawback if such large LPs
have to be solved at every node of the branch-and-bound tree. On the other hand,
only few of these constraints are binding at the optimum vertex and most of them are
(depending on the objective function) actually not needed at all. The cutting plane
approach makes use of this fact and tries to incorporate only important inequalities.

Suppose that the LP $\max\{c^T x \mid Ax \le b, 0 \le x \le 1\}$ with very many constraints
has to be solved. The *cutting plane approach* solves such an LP as follows.

Cutting plane algorithm

(1) Initialize P as linear program $\max\{c^T x \mid 0 \le x \le 1\}$.
(2) Solve problem P and obtain an optimum solution x^*.
(3) If $Ax^* \le b$, Stop (x^* solves $\max\{c^T x \mid Ax \le b, 0 \le x \le 1\}$).
(4) Otherwise, choose an inequality $A_i x \le b_i$ from $Ax \le b$ such that
    $A_i x^* > b_i$. Augment P by this inequality and goto (2).

The term "cutting plane method" was chosen because in every iteration the current optimum point is cut off since it is infeasible for the real problem.

In step (4) also more than one violated inequality can be added. In practical
computations this is highly advisable. It is also useful to eliminate constraints from
the LP which are not binding at the current optimum. For the reoptimization in (2)
after the addition of constraints, the dual simplex algorithm is most useful because
it can be started immediately since dual feasibility is still given.

As an example we illustrate the algorithm when solving the LP relaxation of
the LOLIB benchmark problem be75np (here $n = 75$). Table 5.1 shows for every
reoptimization the number of inequalities added, removed resp., the current number
of constraints in the LP and the current optimum value. The bound given by the
relaxation is 790989 (the optimum linear ordering has value 790966). Instead of the
possible 39200 only 1781 3-dicycle inequalites were generated.

Table 5.1 Solution of `be75np` with cutting planes

#added ineq.	#removed ineq.	#current ineq.	Objective value
0	0	0	816602
200	0	200	810126
200	33	367	806134
200	33	534	802379
200	70	664	800896
200	55	809	795674
200	51	958	794413
200	83	1075	791815
200	48	1227	791196
126	71	1282	791036
27	35	1274	790990
28	17	1285	790989

Further issues such as how to select inequalities to be added, will be addressed later. Note the fact that, actually, in steps (3) and (4), it is not required to have an explicit list of the constraints but just to be able to answer the question of whether x^* is feasible and, if not, to provide a violated inequality. Furthermore, and this is what makes this approach even more powerful: in principle, we do not even have to know the system of inequalities explicitly. We want to make this more precise.

Let $\max\{c^T x \mid Ax \leq b, x \in \{0,1\}^n\}$ be a formulation of a combinatorial optimization problem as 0/1-IP. From polyhedral theory, it is known that there exists a finite system $Bx \leq d$ such that the vertices of the polyhedron $\{x \mid Bx \leq d\}$ are exactly the feasible 0/1 solutions of $Ax \leq b$. Therefore, if the system is known, then the LOP can be solved as the linear programming problem $\max\{c^T x \mid Bx \leq d\}$.

An inequality $f^T x \leq f_0$ is said to be *valid* with respect to $\{x \mid Ax \leq b, x \in \{0,1\}^n\}$ if $f\bar{x} \leq f_0$ for all $\bar{x} \in \{x \mid Ax \leq b, x \in \{0,1\}^n\}$.

If, in the cutting plane algorithm, we can find violated valid inequalities w.r.t. the set $\{x \mid Ax \leq b, x \in \{0,1\}^n\}$ as long as the current point is not feasible then we could in principle not only solve the relaxation but also the combinatorial optimization problem if we have access to all necessary inequalities from the system $Bx \leq d$. This algorithm would be correct under the assumptions that all occuring LPs can be solved, that a cutting plane can always be generated, and that it terminates after a finite number of iterations.

Therefore, at the core of applying linear programming to combinatorial optimization we have the following problem.

Separation Problem

Given a description $\{x \mid Ax \leq b, x \in \{0,1\}^n\}$ of a combinatorial optimization problem and a vector y^*, either prove that y^* is feasible or find an inequality which is valid for $\{x \mid Ax \leq b, x \in \{0,1\}^n\}$, but violated by y^*.

An algorithm which solves the separation problem is called an *exact separation algorithm*.

Usually, we have the following situation when attempting to design a cutting plane algorithm for a combinatorial optimization problem.

- A linear characterization $\{x \mid Ax \le b, x \in \{0,1\}^n\}$ is available and the feasibility test just amounts to checking whether x^* is binary.
- The system $Bx \le d$ is not known, but some partial information is available and there are types (called *classes*) of inequalities which have similar structure or logic behind them. Classes may (and will) contain many inequalities (usually at least expontially many in n).
- For some of these classes it can be checked in polynomial time whether all inequalities of the class are satisfied and, if not, a cutting plane can be provided. These are classes of inequalities with *exact separation algorithms*.
- As can be expected for NP-hard problems there will be classes of inequalities the separation of which is NP-complete or where no good separation algorithm is known. In such a case we have to be content with *separation heuristics*. But, also if exact separation is possible, heuristics may be employed in addition to save computing time.

It might seem that cutting planes offer a chance to escape the inherent complexity of NP-hard combinatorial optimization problems. In this context the following result (see e.g. [78]) is of major importance.

Theorem 5.1. *Let $P = \{x \mid Ax \le b\}$ be a rational polyhedron such that the encoding length of every inequality in $Ax \le b$ is at most φ.*

Then for every vector $c \in \mathbf{Q}^n$ the optimization problem $\max\{c^T x \mid Ax \le b\}$ can be solved in polynomial time (in φ and the encoding length of c) if and only if for every vector $y \in \mathbf{Q}^n$ the separation problem with respect to P can be solved in polynomial time (in φ and the encoding length of y).

This theorem states the equivalence of optimization and separation. Therefore, LP relaxations with polynomial separation algorithms can be solved in polynomial time. The number of inequalities is not relevant as long as cutting planes can be generated in polynomial time. It should be noted that this result is only true if the respective linear programs are solved using the ellipsoid method. Since this method is not practicable, the simplex algorithm is used in practice. No polynomial version of the simplex method could be formulated so far but, in any case, it is a fast algorithm for practical computations.

We now have a theoretical framework for solving combinatorial optimization problems using linear programming and branch-and-bound. The remaining sections of this chapter will be devoted to practical issues of the implementation and possibilities to add further cuts which are not contained in the LP relaxation and not known beforehand.

5.3 Branch-and-Cut with 3-Dicycle Cuts

A branch-and-bound algorithm using the canonical LP relaxation for computing upper bounds can, in principle, be used for solving the LOP to optimality. An interesting theoretical result should be mentioned here. Recall the Lagrangean relaxation with 3-dicycle inequalities of Chap. 4:

$$L(\mu) = \max \sum_{(i,j)\in A_n} c_{ij}x_{ij} + \sum_{(r,s,t)\in T} (2 - x_{rs} - x_{st} - x_{tr})\mu_{rst}$$
$$x_{ij} + x_{ji} = 1,\ 1 \le i < j \le n$$
$$x_{ij} \in \{0,1\},\ 1 \le i, j \le n, i \ne j.$$

Because the constraints of the Lagrangean problem define a polyhedron with 0/1 vertices only, it can be shown that the optimum value $\min_{\mu \ge 0} L(\mu)$ is equal to the optimum value of the canonical LP relaxation of the LOP.

We now address the question of how to solve the LP relaxation most effectively and we will see that its exact value can be obtained much faster than the approximate value using the Lagrangean approach.

5.3.1 Solving the 3-Diycle Relaxation

The 3-dicycle relaxation to be solved at every node of the tree is

$$\max \sum_{(i,j)\in A_n} c_{ij}x_{ij}$$
$$x_{ij} + x_{jk} + x_{ki} \le 2,\ \text{for all distinct nodes } i,\, j,\, k \in V_n$$
$$x_{ij} + x_{ji} = 1,\ 1 \le i < j \le n$$
$$x_{ij} \ge 0,\ 1 \le i, j \le n, i \ne j.$$

In the branching process some of the variables will be fixed.

When looking for violated 3-dicycle inequalities it makes no sense to look for some sophisticated procedure: it is appropriate to just enumerate all of them and check for violation. But there are some possibilities to speed up the solution process.

An important observation is that it is helpful to try to generate "deeper" cuts. To this end we set z as a convex combination of the current LP optimum x^* and the currently best known feasible solution $\bar{x}$ and first try to cut off z. Only if z cannot be cut off, then 3-dicycle inequalities separating x^* are searched for. Experiments showed that the setting $z = \frac{1}{3}x^* + \frac{2}{3}\bar{x}$ leads to good results and usually reduces the overall CPU time by about 60%.

Furthermore, it is not necessary to add all violated inequalities in every phase, but to limit their number. Significant progress in the first phases is already achieved with few inequalities and thus the LP size only increases slowly. The best limit on the number of added cuts depends on the problem size, but usually limiting the number to about several hundreds is perfect. It is a further improvement to add in the first phases only inequalities which are arc-disjoint, i.e., having the property that no variable occurs in more than one cut.

It is not only important to limit the number of cuts added, but also to select among the available cuts. This is substantiated in Table 5.2. Columns 5 and 6 show the number of LPs solved (nlps$_2$) and the CPU time (CPU$_2$) in min:sec if cuts were just selected at random. Alternatively, cuts $f^T x \leq f_0$ with large value $\frac{c^T f}{\|f\|}$ were preferred. The idea is that inqualities with small angle between their normal vector and the normal vector of the objective function should yield progress. The results for this strategy are shown in columns 3 and 4 (nlps$_1$ and CPU$_1$). Depending on the problem it can lead to substantial savings in CPU time (see in particular atp76).

Table 5.2 Importance of cut selection strategies

Problem	Size	nlps$_1$	CPU$_1$	nlps$_2$	CPU$_2$
EX1	50	15838	7:15	18065	8:38
EX2	50	20993	10:26	24977	13:07
EX4	50	244	0:06	365	0:14
EX5	50	346	0:10	445	0:15
EX6	50	138	0:04	203	0:08
atp66	66	112	0:06	278	0:29
atp76	76	591	0:49	4186	14:28
econ77	77	30	0:02	63	0:04
randD	50	9711	4:14	10097	4:20
Sum		48003	23:16	58679	41:46

5.3.2 An LP Based Heuristic

The following hypothesis proved to be very useful. We assume that the current fractional LP solution x^* should be somehow close to an optimum solution and contain some of its characteristics. Therefore, we exploit x^* for starting a heuristic to find a feasible solution. For every object i we compute

$$s_i = \sum_{j \neq i} x_{ij}^*$$

and then sort the objects according to nondecreasing values s_i. The corresponding linear ordering serves as a starting solution for improvement heuristics.

For proving the usefulness of this idea, we solved the 3-dicycle relaxation for the LOLIB input-output matrices and called the lower bound heuristic for every LP solution. As improvement heuristic we just employed local enumeration as described in Chap. 2.

Since almost all problems could be solved to optimality we list in Table 5.3 only those problems that could not be solved and the artifical problems `econ36` `-econ77`. The table shows in column 3 the deviation of the 3-dicycle upper bound from the optimum solution in percent. For these easy problems the bound is very good and at most 0.052% above the optimum. (The bound for `stabu70` is less than 0.0005% off).

In addition, Table 5.3 proves that the LP based heuristic works very well achieving at least 99.26% of the optimum.

Table 5.3 3-Dicycle relaxation

Problem	OPT	3CYC	LPHeu
be75np	790966	0.003%	100.000%
stabu70	422088	0.000%	99.938%
t59b11xx	245750	0.005%	100.000%
econ36	555568	0.003%	99.888%
econ43	675180	0.001%	99.670%
econ47	845374	0.032%	99.889%
econ58	1263005	0.004%	99.358%
econ59	1256708	0.035%	99.726%
econ61	1265164	0.002%	99.881%
econ62	1275989	0.005%	99.263%
econ64	1328547	0.011%	99.627%
econ67	1437471	0.007%	99.969%
econ68	1480971	0.052%	99.770%
econ71	1636218	0.007%	99.624%
econ72	1932752	0.028%	99.795%
econ73	2146505	0.009%	99.648%
econ76	2781838	0.008%	99.735%
econ77	2798507	0.007%	99.994%

5.3.3 Computational Results with 3-Dicycles

In another experiment we solved the same instances as with the branch-and-bound algorithm of Charon and Hudry of Chap. 4. Table 5.4 displays for every problem the root bound obtained from the 3-dicycle relaxation, the number of nodes in the branch-and-cut tree, the maximum depth of the tree, and the CPU time (in min:sec).

Table 5.4 Branch-and-cut with 3-dicycle inequalities

Problem	Optimum	Root bound	#nodes	#maxlevel	CPU
p40−01	29457	29494.47	3	2	0:00
p40−02	27482	28032.00	5449	28	15:24
p40−03	28061	28354.33	115	12	0:20
p40−04	28740	29298.67	2317	22	7:12
p40−05	27450	28213.33	976	27	9:26
p40−06	29164	29632.33	613	15	1:54
p40−07	28379	29006.00	2777	21	8:30
p40−08	28267	28870.67	1279	16	4:08
p40−09	30578	31183.00	5675	23	17:00
p40−10	31737	32147.67	419	15	1:07
p40−11	30658	31275.00	3985	23	11:43
p40−12	30986	31479.00	571	14	1:48
p40−13	33903	34056.85	10	5	0:00
p40−14	34078	34494.33	241	13	0:39
p40−15	34659	35369.67	5073	28	14:24
p40−16	36044	36199.00	17	5	0:00
p40−17	38201	38217.86	3	2	0:00
p40−18	37562	37694.93	17	5	0:00
p40−19	38956	39117.17	15	5	0:00
p40−20	39658	39812.67	31	7	0:00

The problems are solved in about the same order of magnitude of CPU time. The branch-and-cut algorithm generates much fewer tree nodes, but solving the LPs needs a lot of time. Of course, since the 3-dicycle bound and exact Lagarangean bound are equal, problems which are difficult for one code are difficult for the other one as well. Recent experiments with a bundle method, however, were very promising. Running some iterations of the bundle method on the Lagrangean relaxation seems to offer the chance of identifying important starting inequalities for computing the 3-dicycle bound with a subsequent cutting plane algorithm.

There have also been experiments using interior-point solvers for optimization the LP relaxations. Restarting after adding cuts is difficult here. In [126, 127] restart is realized by backtracking some steps on the path taken when optimizing the previous LP. Results, however, do not suggest employing interior-points methods in this context.

5.4 Generation of Further Cuts

Only 3-dicycle cuts are not sufficient for attempting to solve larger problems. We will now address the question of how to generate further cuts. As further cuts we will only discuss at this point cuts that can basically be used for every combinatorial optimization problem. Special further cuts for the LOP will be described in Chap. 6.

5.4.1 Chvátal-Gomory Cuts

We discuss briefly the principle of general cuts. Their generation has far-reaching consequences at least in theory, but to some extent also in practice.

Chvátal-Gomory cuts can be viewed as being obtained by so-called closure operations which allow for generating new stronger inequalities from an inequality system.

Definition 5.1. Let S be the system $Ax \leq b$ of rational inequalities.

(i) An inequality $d^T x \leq d_0$ with integral d is said to belong to the *elementary closure* of S if there is a rational vector $\lambda \geq 0$ such that $\lambda^T A = d^T$ and $\lfloor \lambda^T b \rfloor \leq d_0$.
(ii) The set of all inequalities belonging to the elementary closure of S is denoted by $e^1(S)$. For $k > 1$ the set $e^k(S)$ is defined as $e^k(S) = e^1(S \cup e^{k-1}(S))$.
(iii) The *closure* $cl(S)$ of S is given as

$$cl(S) = \cup_{k=1}^{\infty} e^k(S).$$

An inequality is said to have *Chvátal rank* k with respect to S if it is contained in $e^k(S)$, but not in $e^{k-1}(S)$. The number k is a good indicator for the complexity of an inequality.

As an example, assume that for integral variables y and x_i, $i = 1, \ldots, k$ the following constraints have to be satisfied:

$$a_1 x_1 + \ldots + a_k x_k - y = b \tag{5.1}$$
$$-x_i \leq 0, \quad i = 1, \ldots, k. \tag{5.2}$$

By adding suitable multiples of (5.2) to (5.1) and rounding down the right hand side one obtains the valid inequality

$$\lfloor a_1 \rfloor x_1 + \ldots + \lfloor a_k \rfloor x_k - y \leq \lfloor b \rfloor. \tag{5.3}$$

Subtracting (5.1) from (5.3) now yields the inequality

$$(\lfloor a_1 \rfloor - a_1) x_1 + \ldots + (\lfloor a_k \rfloor - a_k) x_k \leq \lfloor b \rfloor - b.$$

Note that this inequality is violated if $x_1 = \ldots = x_k = 0$. This is the standard type of a Chvátal-Gomory cut as it can be obtained directly from the simplex tableau. There the variables $x_1, \ldots, x_n$ are the non-basic variables at the optimum of the current LP relaxation and thus the inequality cuts off this optimum, but is satisfied by all feasible integral points.

A very nice theoretical results is

Theorem 5.2. *A cutting plane algorithm with Chvátal-Gomory cuts either solves an integer programming problem* $\max\{c^T x \mid Ax = b, x \geq 0, x \text{ integer}\}$ *in finitely many steps or verifies that it is unbounded or infeasible.*

For proving this theorem several technicalities have to be observed. The optimal tableau of the LP relaxations has to be lexicographically positive and thus the pivoting rule of the dual simplex has to be modified to preserve lexicographic positivity. A lower bound on a feasible solution (derivable from the data) has to be tested. A cut has to be derived from the row with smallest index. If in the dual simplex the slack variable of a cut becomes basic, then the cut is eliminated from the problem.

It turned out that this approach cannot be utilized in practice straightaway because after the addition of many such cutting planes severe numerical problems occur. On the other hand, however, it was shown that the careful use of Chvátal-Gomory cutting planes can lead to substantial improvements in integer programming.

5.4.2 Maximally Violated Mod-k Cuts

As a further possiblity to generate cuts of general nature in a branch-and-cut algorithm we describe mod-k inequalities which are special kinds of Chvátal-Gomory cuts.

Let $Ax \leq b$ be a system of linear inequalities with integral coefficients and let $k > 1$ be an integer number. Suppose that we scale every inequality r of $Ax \leq b$ by a nonnegative factor μ_r and sum the resulting inequalities. Now, if μ denotes the vector of all factors, assume that all coefficients of $\mu^T A$ are divisible by k and that the remainder on dividing $\mu^T b$ by k is $k-1$. Hence μ satisfies the congruence system

$$\mu^T A \equiv 0^T \mod k$$
$$\mu^T b \equiv k-1 \mod k.$$

We have $\mu^T b = sk + (k-1)$ for some $s \in \mathbf{Z}$, and therefore $\mu^T b - (k-1)$ is divisible by k. Furthermore $\mu^T Ax$ is divisible by k for all $x \in \{x \in \mathbb{Z}^n \mid Ax \leq b\}$, and therefore the inequality

$$\mu^T Ax \leq \mu^T b - (k-1)$$

is valid for all feasible integer solutions of $Ax \leq b$. We can express the inequality in an equivalent way as the *mod-k inequality*

$$\frac{1}{k}\mu^T Ax \leq \frac{1}{k}(\mu^T b - (k-1)).$$

Now consider some fractional solution x^* of $Ax \leq b$. In a branch-and-cut algorithm we would like to find an integer k and a vector μ such that the conguence system above is satisfied and the corresponding mod-k inequality is violated by x^* and thus provides a cutting plane (a so-called *mod-k cut*). Since $\mu^T Ax^* \leq \mu^T b$ this solution can violate $\mu^T Ax \leq \mu^T b - (k-1)$ by at most $k-1$ and the maximal violation can only be achieved if $\mu^T Ax^* = \mu^T b$, i.e., if $\mu_r = 0$ for all r with $A_r.x^* < b_r$.

We use the separation algorithm for maximally violated mod-k cuts suggested in [28]. Let x^* be a fractional solution of the current LP relaxation $Ax \leq b$. Due to the remarks above, in order to find a maximally violated inequality, we restrict the congruence system to contain only those inequalities that are tight for x^*. We have to choose k and find an integer multiplier vector $\mu \geq 0$ such that the congruence system is satisfied. Note that the coefficients of μ can obviously be restricted to values smaller than k, i.e., $\mu_r \in \{0, 1, \ldots, k-1\}$ for all r. It was proved in [28], that only prime numbers have to be considered for k.

Usually, if one cut exists then there are plenty of others. This is due to the fact, that there is a number f of free variables μ_i in the solution whose values can be chosen arbitrarily from $\{0, 1, \ldots, k-1\}$. There can be dozens or even a few hundred of these variables and thus there exist k^f different solutions of the congurence system. Even though the map of solutions to cuts is not injective, we will have to address the problem of selecting cuts from the huge set of generated constraints to be added to the linear relaxation.

5.4.3 Mod-2 Cuts

For the special case $k = 2$ an efficient separation procedure is given in [27]. Again, we search for a multiplier vector μ for $Ax \leq b$ that satisfies the congruence system. The arguments of previous section show that

$$\frac{1}{2}\mu^T Ax \leq \frac{1}{2}(\mu^T b - 1). \tag{5.4}$$

is valid for $\{x \in \mathbb{Z}^n \mid Ax \leq b\}$ and can be violated by at most $\frac{1}{2}$ by a tight fractional solution x^* of $Ax \leq b$. For $\lambda = \frac{1}{2}\mu$ and μ satisfying the congruence system, all entries of $\lambda^T A$ are integer and $\lambda^T b - \frac{1}{2} = \lfloor \lambda^T b \rfloor$. Therefore (5.4) can be expressed equivalently as

$$\lambda^T Ax \leq \lfloor \lambda^T b \rfloor.$$

This inequality is a so-called $\{0, \frac{1}{2}\}$-Chvátal-Gomory inequality. The identification of a suitable vector λ in principle amounts to optimizing over

$$P_{1/2} = \left\{ x \in \mathbb{R}^n \mid Ax \leq b, \lambda^T Ax \leq \lfloor \lambda^T b \rfloor \text{ and } \lambda \in \{0, \tfrac{1}{2}\}^m \text{ with } \lambda^T A \in \mathbb{Z}^n \right\}.$$

As the separation problem of $P_{1/2}$ is NP-hard, it is unlikely that one can optimize over $P_{1/2}$ in polynomial time. But, in [27], possibilities are derived for relaxing $P_{1/2}$ in such a way that separation becomes polynomially solvable.

By a so-called *weaking of inequalities*, a separation algorithm is developed which basically amounts to solving shortest path problems in a specially constructed digraph.

The advantage of this *shortest path mod-2 method* is that there is no restriction on the constraints that are used for the generation of the cut, as long as they can be transformed in such a way that they have exactly two odd coefficients on the left hand side. Another advantage is that the digraph for the shortest path computations is usually sparse. On the other hand, no extension from 2 to bigger prime numbers is possible and the violation of the resulting cut is maximal if and only if all used constraints are tight for x^*.

There is a further problem-independent approach for generating cuts, namely so-called *local cuts* and *target cuts*. Because it is easier to explain these cuts in the context of the linear ordering polytope, their discussion is postponed to Sect. 6.7.

5.5 Implementation of Branch-and-Cut

We will now discuss in more detail implementational issues for the realization of a branch-and-bound algorithm where upper bounds are computed using LP relaxations. Because the LP relaxations are solved with the cutting plane approach the notion "branch-and-cut" was coined for this type of algorithms.

We will keep the discussion on a general level valid for all combinatorial optimzation problems. At some point we will add special remarks for the LOP.

In the following it is assumed that the problem is defined on a graph or digraph and that the variables are associated with edges or arcs. Variables and arcs/edges are used as synonyms.

A 0/1-IP formulation of the problem is assumed to be available as

$$\max\{c^T x \mid Ax \leq b, x \in \{0,1\}^n\}.$$

The minimal equation system is also known, but for simplicity not listed explicitly. Further inequalities $Ax \leq b$ in addition to the system $\tilde{A}x \leq \tilde{b}$ are known with

$$\{x \mid Ax \leq b, x \in \{0,1\}^n\} \subset \{x \mid \tilde{A}x \leq \tilde{b}\} \subset \{x \mid Ax \leq b, 0 \leq x \leq 1\}.$$

Note that an explicit list of the constraints of $\tilde{A}x \leq \tilde{b}$ is not needed. We assume that for some classes of inequalities in $\tilde{A}x \leq \tilde{b}$ exact or heuristic separation algorithms are available.

The algorithm will construct a branch-and-cut tree whose nodes represent the subproblems generated. The following states of a tree node will be distinguished:

- *current*: this node is presently worked on,
- *active*: the node is generated, but not considered yet,
- *inactive*: the node has been treated and has active successors,
- *fathomed*: this node and all of its successors have been worked on (such nodes can obviously be deleted from the tree).

A variable can be in one of the following states:

- *active*: the variable is present in the current LP,
- *inactive*: the variable is not present in the current LP,

  – *fixed*: the variable is permanently fixed to 1 or 0,
  – *set*: the variable is set to 1 or 0, but this is valid only in part of the tree.

Furthermore, let L be the global lower bound on the optimum objective function
value (best known feasible solution so far) and let U be the global upper bound
(valid for the original problem).

  We now describe the main aspects of the branch-and-cut algorithm. The goal is
to solve $\max\{c^T x \mid Ax \leq b, x \in \{0,1\}^n\}$ based on the approximation $\tilde{A}x \leq \tilde{b}$.

5.5.1 Initialization

As first LP for the root node $\max\{c^T x \mid 0 \leq x \leq 1\}$ is usually chosen. In principle, the
minimal equation system could be used to eliminate variables beforehand. However,
this is only reasonable in few cases, such as the LOP. Here variables x_{ij}, $i > j$, can
simply be replaced by $1 - x_{ji}$ without effecting the density of the constraint matrix.

  One could already introduce some of the inequalities from $Ax \leq b$ if respective
information is available. Computational experience has shown that this only has a
significant effect if inequalities are selected carefully.

5.5.2 Active Variables

If the number of variables is large, it can be useful to work only with a subset of
active variables and assume that the inactive variables are zero. Of course, it has to
be checked later (by pricing) if this assumption is correct.

  For the LOP, working with only a subset of variables has not been interesting so
far. The problem is already difficult for small values of n, and the number $n(n-1)$
of its variables ($\binom{n}{2}$ after elimination, resp.) is still small.

5.5.3 Local Upper Bound

The solution of the current LP gives an upper bound for the respective subproblem
and all subproblems generated from it (if all inactive variables price out correctly).
LPs are usually solved using the dual simplex method because of the simple restart
after addition of constraints or variables.

5.5.4 Branching

If the current subproblem cannot be solved, it has to be partitioned. A common way for partitioning (also frequently used for the LOP) is to choose some variable x_{ij} with fractional value and set it to 1 in one subproblem and to 0 in the other. This way a binary branch-and-cut tree is generated.

Several priority rules for selecting this branching variable are possible. It is common to choose variables close to $\frac{1}{2}$ with high absolute value of their objective function coefficient. The motivation behind this is that setting the branching variable should have some effect on the current problem. Non-binary partitions or generating subproblems using inequalities is also possible.

5.5.5 Fixing and Setting of Variables

Let x^* be the optimum solution of the current LP with value c^* and let r be the vector of reduced costs.
The following holds for a nonbasic variable x_e:

  (i) If $x_e = 0$ and $c^* + r_e \leq L$, then one can set $x_e = 0$.
  (ii) If $x_e = 1$ and $c^* - r_e \leq L$, then x_e can be set to 1.

This setting of variables is valid for the current node and all of its successors.

If setting is possible at the root node, then it amounts to fixing a variable permanently. If advantageous, the variable could be removed from the problem.

5.5.6 Logical Implications

Setting variables can (depending on the problem) influence other variables as well. E.g. if in the LOP we have the setting $x_{ij} = 1$ and $x_{jk} = 1$, then $x_{ki} = 0$ is implied, because otherwise a 3-dicycle inequality would be violated.

Tests for logical implications can in particular be helpful since it can affect basic variables which are currently fractional. Nonbasic variables are only set or fixed to values they already have (so, locally there is no effect on the problem).

5.5.7 Selection of Nodes

When work on the current node is finished, then the next node for processing has to be chosen. Basic strategies are:

  – *Depth-first*
    Choose an active node with highest level in the branch-and-cut tree.

- *Breadth-first*
  Choose an active node with lowest level in the branch-and-cut tree.
- *Best-first*
  Choose a node with smallest difference between its lower and upper bound. (Note that different upper bounds are valid for the nodes.)
- *Strong branching*
  Here work is started on several active nodes and depending on the progress one of them is selected promising the best improvement of the upper bound.

Depth-first is mainly employed if it is at all difficult to find feasible solutions. Otherwise it is inferior compared with the other strategies. Best-first leads to fewer nodes and better running times than breadth-first. Strong branching usually performs best, but at higher effort.

Further heuristics rules for selecting the most promising node can be formulated in addition.

5.5.8 Lower Bounds

Good feasible solutions are very important in order to obtain lower and upper bounds which are close together. Feasible solutions can be computed independent from the branch-and-cut algorithm beforehand or in parallel on a separate processor. As pointed out above, we found LP based heuristics exploiting the current fractional LP solution very useful.

Note that, if better lower bounds are found, fixing further variables can be tried. To this end the reduced costs at the root node have to be stored.

5.5.9 Separation

Separation clearly is at the core of branch-and-cut. It depends on the knowledge about the system $\tilde{A}x \leq \tilde{b}$ and on how many effective separation algorithms are available. Strategies for calling the respective procedures and for the selection of cutting planes found have to be developed.

5.5.10 Elimination of Constraints

With the cutting plane approach only a very small fraction of potential inequalities is actually introduced into the LP, but nevertheless this number also can be huge. It has proved to be reasonable to eliminate inequalities which are not binding at the current LP optimum. However, these inequalities should be stored in a constraint

pool, because they could be valuable for other nodes. This is in particular advisable
if they have been found with high computational effort.

5.5.11 Constraint Pool

The constraint pool stores inequalities that are needed for the initialization of the
next node or were temporarily eliminated. Since the pool can grow very large, suitable data structures for storing inequalities in sparse format are necessary.

5.5.12 Pricing

If the current LP is solved in the active variables then it has to be checked if the
setting of the inactive variables is optimal as well. This can be accomplished by
evaluating their reduced costs.

Assume that x_e is an inactive variable with value 0. First the column for x_e in the
constraint matrix has to be retrieved; let this be a_e. Then the equation system

$$A_B \bar{a}_e = a_e$$

has to be solved. The reduced costs of x_e are

$$r_e = c_e - c_B^T \bar{a}_e = c_e - c_B^T A_B^{-1} a_e.$$

Usually, in LP codes, the dual vector $c_B^T A_B^{-1}$ is available and does not have to be
computed. So there is no need for solving an equation system.

If $r_e \geq 0$, for all inactive variables x_e, then the current LP solution is optimal for
the complete problem, although possibly only a small subset of the variables has
actually been used. If the reduced costs are negative for one or more variables, then
some of them have to be activated and the LP has to be augmented correspondingly.

It is possible to perform pricing in several stages by assigning priority classes
to inactive variables. Also, not all variables will be examined, if already several
have been found that do not price out correctly. Note that pricing incurs significant
computational effort.

5.5.13 Infeasible LPs

If there are inactive variables it is possible that the current LP is infeasible. This
can be caused by a setting of variables that makes it impossible to achieve feasibility with active variables only. In such a case variables have to be added to regain

feasibility. It is not clear how to choose such variables, and there are only heuristic strategies available.

5.5.14 Addition of Variables

Addition of variables is necessary if LPs become infeasible or it can also be caused by pricing. It can also be helpful to activate variables which seem to be important because they are contained in good feasible solutions.

The efficient implementation of the above components is not trivial. In particular, the administration of branch-and-cut nodes, of active variables and of the constraint pool require some effort. But these are tasks which can be solved to a large extent independently of the concrete problem and there are frameworks like ABACUS [93], SCIP [2] or SYMPHONY [141] that facilitate the development of branch-and-cut algorithms. Cutting plane generation is problem dependent in any case and also requires complex algorithms and data structure.

5.6 Some Computational Results

We report about computational results from [135] where also further details on the implementation of mod-2 and mod-k separation can be found.

We have applied the branch-and-cut algorithm with mod-2 and mod-k cuts on ten random problem instances p40-01 - p40-10. The problem instances are difficult and no 3-dicycle relaxation had an integral optimum solution. In the following we describe how we deal with the fractional solution x^* of the relaxation. We will also speak about the digraph associated with x^* which consists of the arcs whose associated variables have a positive value.

Since the generation of cuts is time consuming, we incorporated a heuristic element in our separation procedures to possibly generate further cuts at cheap cost. Our idea is based on an interesting property of the linear ordering polytope which will be discussed in the next chapter. Namely, for this polytope a rotation mapping can be defined which converts valid inequalities into different valid inequalites. We enhanced our separation routines by also checking rotated versions of violated inequalities for violation. This lead to the detection of further cuts.

We are using two general strategies for trying to generate mod-k cuts violated by the current fractional solution x^*. The first strategy considers small subdigraphs (of the digraph defined by x^*) and generates all violated cuts that can be found for this digraph. The idea is that by proceeding this way, the relaxation can locally be strengthed considerably and therefore allow for a reasonable bound improvement in the branch-and-cut algorithm. The second strategy applies the mod-k separation routine to the complete digraph and selects cuts afterwards. Cuts for the whole digraph should provide global information which is also important for the algorithm.

The strategies for choosing an appropriate subdigraph differ as follows. The *variable heuristic* limits the number n_V of variables from which the subdigraph is constructed, while the *improvement heuristic* limits the number n_N of the nodes of the subdigraph.

The variable heuristic starts with the variable that occurs in the most constraints of the current LP. (If there are several variables satisfying this condition, then we choose one at random.) Then we continue to successively select all other variables from these constraints and also choose all constraints that contain these new variables. If the limit n_V on the number of variables is reached, the selection is stopped and the subdigraph G' is defined by the variables selected by this procedure.

A practical problem we had to address was that already a slight increase of n_V could lead from subdigraphs for which no cuts were found to subdigraphs where very many violated cuts existed. This phenomenon turned out to be caused by the fact that not all 3-dicycle inequalities for the subdigraph are part of the LP relaxation. After all trivial and binding 3-dicycle inequalities were added, this problem disappeared. We experienced that, for our problem instances, $n_V = \frac{1}{4}n$ and $n_V = \frac{1}{5}n$ are good bounds as the resulting digraph was just big enough to generate a reasonable number of mod-k cuts.

The improvement heuristic uses information from the current LP solution. For every node, the sum of the values of variables corresponding to outgoing arcs is computed. Then the nodes are linearly ordered with respect to nonincreasing sums. Our heuristic is based on the hypothesis that nodes close together in this ordering will also be neighboring in an optimum solution. For forming the subdigraph we therefore take the first n_N nodes of the ordering, then the second n_N nodes, etc., and generate all mod-k cuts from these n/n_N subgraphs. Values $n_N \in \{\frac{1}{4}n, \frac{1}{5}n, \frac{1}{6}n\}$ lead to suitable results.

The third strategy is to apply the maximally violated mod-k method to the whole digraph and select cuts afterwards. This selection strategy is based on some different criteria. First, we randomly order the columns of the matrix to avoid always generating the first cuts obtained from the system in each iteration. Second, we prefer cuts introducing few non-zero coefficients because dense LPs are usually more difficult than sparse ones. A further criterion is to select only the single basic solution or at most f trivial solutions and not take all k^f possible solutions of the system into account.

For the shortest path calculation of the mod-2 procedure we use Dijkstra's algorithm leading to running time $O(\overline{E} \log \overline{V})$. To avoid inequalities being found more than once we delete all nodes that are part of an already found violated inequality from the list of potential starting nodes.

In the case of rotation we repeat this procedure for all rotation parameters r in a random order until the limit of 250 cuts is reached. For the violation tolerance we use the relatively high value of 0.01, because otherwise the separation procedure finds a lot of cuts without significantly improving the dual bound. In addition we use tailing off which stops the mod-2 separation if the last 10 separations did not improve the bound by more than 0.05 percent.

Table 5.5 Root bounds

Problem	Opt	3-cyc	MV	M2	MV+M2
p40-01	29457	29494.48	29488.09	29457.00	29457.00
p40-02	27482	28032.00	27967.75	27768.25	27768.11
p40-03	28061	28354.33	28297.53	28098.77	28098.59
p40-04	28740	29298.67	29220.46	28976.85	28974.93
p40-05	27450	28213.33	28091.60	27855.72	27849.33
p40-06	29164	29632.30	29553.90	29321.40	29316.30
p40-07	28379	29006.00	28892.08	28651.53	28647.43
p40-08	28267	28870.67	28744.77	28463.46	28460.07
p40-09	30578	31183.00	31078.85	30898.49	30891.87
p40-10	31737	32147.67	32089.26	31897.68	31894.69

Table 5.5 displays, for the instances p40-01 – p40-10, the optimum values, the 3-dicycle bound and the improved bound when additional cuts are added for tightening the LP relaxation. In the first two experiments we added mod-2 and mod-k cuts separately (MV and M2), in the third experiment both separations were employed (MV+M2). The branch-and-cut algorithm was stopped when no more violated inequalities could be found at the root node, i.e., no branching was started.

Table 5.6 verifies that these bounds can be improved considerably with little additional computational effort when rotation is employed to find further cuts.

Improvement with respect to the 3-dicycle relaxation is easier to assess if the *gap closure*

$$100 \times \frac{|c^T x^* - c_{3\text{cycle}}|}{|c_{\text{opt}} - c_{3\text{cycle}}|}$$

is computed, where c_{opt}, $c_{3\text{cycle}}$, and $c^T x^*$ are the optimum objective function value, the 3-dicycle upper bound, and the bound obtained with additional mod-k cuts, respectively. The gap closure measures (in percent) how much of the gap between 3-dicycle bound and optimum value could be closed.

Table 5.7 shows that the gap between 3-dicycle bound and optimum can be closed by 86% on average and that the gap closure is mainly due to mod-2 cuts with rotation.

So with respect to bound improvement the additional separation has proved its advantages. But it is interesting as well to check if this improvement also leads to faster computation times when a provably optimum solution has to be computed.

Our experiments revealed that the separation of maximally violated mod-k inequalities did not have a too big effect on the root bound and is, in general, not worth the effort compared to mod-2 separation. Mod-2- separation gives a better gap closure, and because of the better bounds the number of branch-and-nodes needed to solve the problems to optimality is considerably less (even when rotation is not

Table 5.6 Root bounds with rotation

Problem	Opt	3-cyc	MV_R	$M2_R$	$MV+M2_R$
p40-01	29457	29494.48	29485.69	29457.00	29457.00
p40-02	27482	28032.00	27956.53	27648.32	27647.84
p40-03	28061	28354.33	28290.45	28061.00	28061.00
p40-04	28740	29298.67	29193.68	28829.76	28829.77
p40-05	27450	28213.33	28085.01	27679.71	27678.77
p40-06	29164	29632.30	29546.30	29186.20	29186.10
p40-07	28379	29006.00	28879.06	28481.02	28480.75
p40-08	28267	28870.67	28735.02	28311.82	28312.41
p40-09	30578	31183.00	31077.59	30697.73	30698.30
p40-10	31737	32147.67	32090.65	31778.55	31778.37

Table 5.7 Gap Closure

Name	MV	MV_R	M2	$M2_R$	MV+M2	$MV+M2_R$
p40-01	17.05%	23.45%	100.00%	100.00%	100.00%	100.00%
p40-02	11.68%	13.72%	47.96%	69.76%	47.98%	69.85%
p40-03	19.36%	21.18%	87.12%	100.00%	87.19%	100.00%
p40-04	14.00%	18.79%	57.61%	83.95%	57.95%	83.94%
p40-05	15.95%	16.81%	46.85%	69.99%	47.69%	70.03%
p40-06	16.74%	18.36%	66.39%	95.26%	67.48%	95.28%
p40-07	18.17%	20.25%	56.53%	83.73%	57.19%	83.77%
p40-08	20.86%	22.47%	67.46%	92.58%	68.02%	92.48%
p40-09	17.21%	17.42%	47.03%	80.21%	48.23%	80.01%
p40-10	14.22%	13.88%	60.87%	89.88%	61.60%	89.93%

invoked). Therefore, we did not use mod-k separation anymore. Table 5.8 displays the respective results of our computations.

Table 5.8 CPU times (min:sec), number of tree nodes and percentage of separation time

	3-cyc			M2			$M2_R$		
Name	CPU	#nod	sep	CPU	#nod	sep	CPU	#nod	sep
p40-01	0:00	3	30%	0:01	1	70%	0:01	1	75%
p40-02	6:57	5441	4%	26:26	147	87%	34:51	41	91%
p40-03	0:08	115	6%	0:32	5	79%	0:23	1	79%
p40-04	3:18	2317	4%	12:48	69	86%	13:56	19	89%
p40-05	1:41:36	24317	3%	1:29:39	471	85%	1:53:12	73	93%
p40-06	0:48	609	4%	3:02	19	84%	2:03	3	84%
p40-07	4:22	2775	3%	15:47	85	87%	12:05	15	87%
p40-08	1:57	1287	4%	6:21	33	85%	4:45	3	87%
p40-09	7:42	5625	4%	31:57	177	87%	19:41	23	89%
p40-10	0:33	421	4%	3:14	19	86%	3:41	7	87%

Table 5.8 exhibits the following facts. If only 3-dicycle separation is used, the algorithm spends most of its CPU time for solving the linear programs. Mod-2 separa-

tion changes this relation. Now separation is responsible for the CPU time. Because of the better bounds, the number of branch-and-cut nodes is drastically decreased. However, since separation is time consuming, the overall solution time for the problems with $n = 40$ is not reduced. But this changes for larger problems. E.g. instance p50-01 can now be solved in half of the time.

Our computational experiments for the linear ordering problem lead us to the conclusion that the incorporation of general cut generation procedures is worthwhile and promising, and should also be tried for other combinatorial optimization problems. The optimal use of mod-2 and mod-k inequalities still has to be explored. To some extent their potential cannot be fully exploited because LPs are becoming more difficult, at least for current LP solvers.

Chapter 6
The Linear Ordering Polytope

Abstract So far we developed a general integer programming approach for solving the LOP. It was based on the canonical IP formulation with equations and 3-dicycle inequalities which was then strengthened by generating mod-k-inequalities as cutting planes. In this chapter we will add further ingredients by looking for problem-specific inequalities. To this end we will study the convex hull of feasible solutions of the LOP: the so-called linear ordering polytope.

6.1 Polyhedral Combinatorics and Basic Results

We recall some necessary definitions and notations. If E is some set we denote by $\mathbf{R}^E$ the set of real vectors having $|E|$ components which are indexed by the members of E. The vector with all components equal to 1 is denoted by $\mathbf{1}$ and the zero vector is denoted by $\mathbf{0}$. The unit matrix is denoted by I and the k-th unit vector by e_k. We do not usually state explicitly the row and column dimensions of a matrix or the dimension of a vector if they are clear from the context.

Let $m > 0$ and $x_1, \ldots, x_m \in \mathbf{R}^n, \alpha_1, \ldots, \alpha_m \in \mathbf{R}$. A linear combination $\sum_{i=1}^m \alpha_i x_i$ is called an *affine combination* if $\sum_{i=1}^m \alpha_i = 1$, a *convex combination* if $\sum_{i=1}^m \alpha_i = 1$ and $\alpha_i \geq 0, i = 1, \ldots, m$, and a *conic combination* if $\alpha_i \geq 0, i = 1, \ldots, m$. If $S \subseteq \mathbf{R}^n$, then the *affine hull* aff(S) of S is defined as the set of all (finite) affine combinations of elements of S. Similarly the *convex hull* conv(S) and the *conic hull* cone(S) are defined. By definition we have aff$(\emptyset) = $ conv$(\emptyset) = \emptyset$. If $S = $ aff(S) then S is called an *affine space*. The *affine rank* arank S of a set S is the smallest cardinality of a set X such that $S \subseteq $ aff(X). If $\mathbf{0} \in $ aff(S) then we have arank $S = $ rank $S + 1$; otherwise the affine rank is equal to the linear rank.

For $a \in \mathbf{R}^n \setminus \{\mathbf{0}\}$ and $a_0 \in \mathbf{R}$ the set $\{x \in \mathbf{R}^n \mid a^T x = a_0\}$ is called a *hyperplane*. A hyperplane defines the *halfspace* $\{x \in \mathbf{R}^n \mid a^T x \leq a_0\}$. A *polyhedron* (or $\mathscr{H}$-*polyhedron*) P is defined as the intersection of finitely many halfspaces or equivalently as the solution set of a finite system of linear inequalities. More precisely, P is an $\mathscr{H}$-polyhedron if there exists a matrix $A \in \mathbf{R}^{m \times n}$ and a vector $b \in \mathbf{R}^m$ such that

© Springer-Verlag GmbH Germany, part of Springer Nature 2022
R. Martí and G. Reinelt, *Exact and Heuristic Methods in Combinatorial Optimization*, Applied Mathematical Sciences 175, https://doi.org/10.1007/978-3-662-64877-3_6

$$P = \{x \mid Ax \leq b\}.$$

We use the abbreviation $P(A,b)$ for the set $\{x \mid Ax \leq b\}$.

A description of a polyhedron by means of linear inequalities is also called the *outer description*. Note that some inequalities may actually be equations. If we want to emphasize that equations are present in the description of P, we explicitly write $P = \{x \mid Ax \leq b, Bx = d\}$.

A set P is a called a $\mathcal{V}$-*polyhedron* if there exist finite sets X and Y such that

$$P = \mathrm{conv}(X) + \mathrm{cone}(Y),$$

i.e., P consists of all vectors $z = x + y$ where $x \in \mathrm{conv}(X)$ and $y \in \mathrm{cone}(Y)$. This type of description is called the *inner description* with *generating sets* X and Y.

According to classical results in polyhedral theory, a set P is a $\mathcal{V}$-polyhedron if and only if it is an $\mathcal{H}$-polyhedron.

A *polytope* is a bounded polyhedron, i.e., $P \subseteq \mathbf{R}^n$ is a polytope if and only if it is a polyhedron and there exist bounds $l, u \in \mathbf{R}^n$ such that $l \leq x \leq u$ for all $x \in P$. In particular, P is a polytope if and only if it is equal to the convex hull of a finite set.

The *dimension* of a set S is defined as $\dim S = \mathrm{arank} S - 1$. (The empty set has dimension -1.) The dimension of a polyhedron P is obtained as follows. If $Bx = d$ is a system of equations such that $\mathrm{aff}(P) = \{x \mid Bx = d\}$, then $\dim P = n - \mathrm{rank} B$, where $\mathrm{rank} B$ is the usual rank of the matrix B. A polyhedron $P \subseteq \mathbf{R}^n$ is *full dimensional* if $\dim P = n$. Therefore, if P is full dimensional, then there exists no equation $a^T x = a_0$, with $a \neq \mathbf{0}$, satisfied by all points in P.

An inequality $a^T x \leq a_0$, $a \neq \mathbf{0}$, is said to be *valid* for a polyhedron P if $P \subseteq \{x \mid a^T x \leq a_0\}$. If $a^T x \leq a_0$ is valid then the set $F = P \cap \{x \mid a^T x = a_0\}$ determines a *face* of P (which may be empty). If $F \neq P$ then F is called a *proper face*. If $|F| = 1$ then the element $v \in F$ is called a *vertex* of P. Vertices cannot be represented as convex combinations of other elements of the polyhedron. If P is a polytope, then $P = \mathrm{conv}(V)$ where V is the set of vertices of P. A maximal nonempty proper face F is called a *facet* of P. A face F is a facet of P if and only if $\dim F = \dim P - 1$. If F is a facet we call $a^T x \leq a_0$ a *facet defining* inequality for P. If $a^T x \leq a_0$ and $b^T x \leq b_0$ are inequalities defining the same facet of P, then one can be obtained from the other by multiplication by a positive constant and by adding multiples of equations valid for P. For a full dimensional polyhedron, facet defining inequalities are unique up to positive multiples.

The following theorem shows two ways for proving that an inequality is facet defining.

Theorem 6.1. *Let P be a polyhedron and $b^T x \leq b_0$ a valid inequality such that $F = \{x \in P \mid b^T x = b_0\}$ is a proper face of P. Let $Dx = d$ be a minimal equation system and $\mathrm{aff}(P) = \{x \mid Dx = d\}$. Then the following statements are equivalent.*

(i) *F is a facet of P.*
(ii) *$\dim F = \dim P - 1$.*
(iii) *If $F \subset \{x \in P \mid c^T x = c_0\}$ where $c^T x \leq c_0$ is a valid inequality for P then there exist $\lambda \in \mathbf{R}^{n-\dim P}$ and $\mu > 0$ such that $c^T = \mu b^T + \lambda^T D$.*

The main idea of polyhedral combinatorics for solving a combinatorial optimization problem $(E, \mathscr{I}, c)$ is to associate a polytope with it as follows. For $I \in \mathscr{I}$ we define the corresponding *characteristic vector* (or *incidence vector*) χ^I by setting

$$\chi_e^I = \begin{cases} 0, & \text{if } e \in I, \\ 1, & \text{otherwise.} \end{cases}$$

The polytope $P_{\mathscr{I}}$ associated with $(E, \mathscr{I}, c)$ is

$$P_{\mathscr{I}} = \text{conv}(\{\chi^I \mid I \in \mathscr{I}\}).$$

For the LOP we define the *linear ordering polytope* P_{LO}^n as the convex hull of the characteristic vectors of the acyclic tournaments in D_n, i.e.,

$$P_{LO}^n = \text{conv}(\{\chi^I \in \{0,1\}^{n(n-1)} \mid T \subset A_n \text{ is an acyclic tournament }\}).$$

Hence the vertices of P_{LO}^n correspond exactly to the linear orderings of n objects.

If P_{LO}^n were explicitly known, then the LOP could be solved as the linear programming problem

$$\max\{c^T x \mid x \in P_{LO}^n\}.$$

However, to be able to apply linear programming techniques, the above definition is useless. P_{LO}^n has to be represented as an $\mathscr{H}$-polyhedron. Therefore, it is the main goal of this chapter to study the linear description of P_{LO}^n. The best such description would be a so-called *minimal linear description* $P_{LO}^n = \{x \mid Ax \le b, Bx = d\}$ where $\text{aff}(P_{LO}^n) = \{x \mid Bx = d\}$ with B of full row rank and where the inequality system $Ax \le b$ contains exactly one defining inequality for every facet of P_{LO}^n.

Some basic general properties of the linear ordering polytope are easily derived.

Theorem 6.2. *Let $n \ge 2$. Then the system*

$$x_{ij} + x_{ji} = 1, \ i, j \in V_n, i < j,$$

is a minimal equation system for P_{LO}^n.

Due to the minimal equation system the dimension of P_{LO}^n is $\binom{n}{2}$. Since P_{LO}^n is not full-dimensional, a facet defining inequality can be represented in different ways. However, the structure of the equation system allows a simple standard representation of inequalities to be defined.

Theorem 6.3. *For every facet of P_{LO}^n there exists an inequality $a^T x \le \alpha$ defining it such that the vector a has nonnegative integral coefficients and the property that for every pair of nodes $i, j \in V_n$ at least one of the coefficients a_{ij} or a_{ji} is equal to zero.*

We can use this observation to define a *normal form* for facet defining inequalities. Namely, every facet can be represented uniquely by an inequality $a^T x \le \alpha$ such that all coefficients a_{ij} are nonnegative coprime integers and $\min\{a_{ij}, a_{ji}\} = 0$ for every pair of nodes $i, j \in V_n$. The set of arcs corresponding to positive coefficients is called the *support* of a.

An important general question is to decide whether or not two facet defining inequalities define the same facet. A sufficient condition for the nonequivalence of two inequalities is given in the following theorem.

Theorem 6.4. *Let $a^T x \leq \alpha$ and $b^T x \leq \beta$ be facet defining inequalities for P_{LO}^n, $n \geq 2$, given in normal form. If there exists an arc $(i,j) \in A_n$ with $a_{ij} > 0$ and $b_{ij} = 0$ (or $b_{ij} > 0$ and $a_{ij} = 0$) then the inequalities define different facets.*

Two useful general properties of facet defining inequalities for P_{LO}^n are stated in the following two lemmas.

Theorem 6.5 (Trivial Lifting Lemma). *Let $a^T x \leq \alpha$ be facet defining for P_{LO}^n, $n \geq 2$. Define the vector $\bar{a} \in \mathbf{R}^{(n+1)n}$ by setting $\bar{a}_{ij} = a_{ij}$ for all $(i,j) \in A_n$, and $\bar{a}_{i,n+1} = \bar{a}_{n+1,i} = 0$, for $i = 1, \ldots, n$. Then $\bar{a}^T x \leq \alpha$ defines a facet of P_{LO}^{n+1}.*

Theorem 6.6 (Arc Reversal Lemma). *Suppose $a^T x \leq \alpha$ is a facet defining inequality for P_{LO}^n, $n \geq 2$. If $b \in \mathbf{R}^{n(n-1)}$ is defined by $b_{ij} = a_{ji}$ for all $(i,j) \in A_n$, then $b^T x \leq \alpha$ is facet defining for P_{LO}^n.*

Therefore the linear ordering polytope P_{LO}^{n+1} inherits all facets from P_{LO}^n in the sense that the added coefficients can just be set to zero. Furthermore, reversing the arcs in the support of a facet defining inequality yields a new facet defining inequality. However, if the reversed digraph is isomorphic to the original support digraph then the new facet basically has the same structure because only the numbering of the nodes is different.

6.2 Facets of the Linear Ordering Polytope

The facial structure of P_{LO}^n has been investigated in many publications [45, 19, 77, 143, 110, 159, 60, 49, 59]. Interestingly, there has been a lot of independent research because the linear ordering problem occurs with different names in several fields. In this section we review some of the results. We usually give no proofs (except for showing some principles) and do not give definitions precisely that are of minor importance for our exposition.

A simple class of inequalities for P_{LO}^n is given by the so-called **trivial inequalities** $0 \leq x_{ij} \leq 1$, for all $(i,j) \in A_n$, which are always valid for polytopes with 0/1-vertices.

Theorem 6.7. *Trivial inequalities define facets of P_{LO}^n for all $(i,j) \in A_n$. No two of the inequalities $x_{ij} \leq 1$ are equivalent. The normal form of inequality $-x_{ij} \leq 0$ is $x_{ji} \leq 1$.*

The minimal equation system and the trivial inequalities are sufficient for describing P_{LO}^2. Further inequalities are needed for $n \geq 3$. Clearly, inequalities excluding dicycles in tournaments should play a central role.

Theorem 6.8. *For every dicycle C of length three in A_n, the inequality $x(C) \leq 2$ defines a facet of P_{LO}^n.*

Inequalities $x(C) \leq |C| - 1$ for longer dicycles are obviously valid for P_{LO}^n. Let $|C| > 3$ and let i, j be a pair of dicycle nodes with $(i, j) \notin C$ and $(j, i) \notin C$. Then C can be partitioned into C_1 and C_2 such that $C_1 \cup \{(i, j)\}$ and $C_2 \cup \{(j, i)\}$ are dicycles. Now $x(C) = x(C_1) + x(C_2) + x_{ij} + x_{ji} \leq |C_1| - 1 + |C_2| - 1 + 1 = |C| - 1$. All dicycle inequalities for dicycles longer than three are thus implied by 3-dicycle inequalites and therefore do not define facets.

The polytope P_{LO}^3 is contained in $\mathbf{R}^6$, but using the equation system it can be projected to $\mathbf{R}^3$. Figure 6.1 shows the projection of P_{LO}^3. The remaining variables are x_{12}, x_{13} and x_{23}. The vertices are labeled by the corresponding linear orderings.

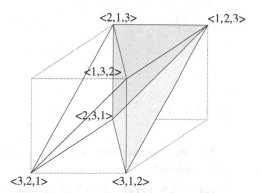

Fig. 6.1 The (projected) polytope P_{LO}^3

The minimal equation system, trivial inequalities and 3-dicycle inequalities are sufficient to completely describe P_{LO}^3, P_{LO}^4, and P_{LO}^5. Hence their respective number of facets is 8, 20, and 40. In earlier publications [45, 19] it was believed that, in general, the polytope P_C^n defined by equations, trivial and 3-dicycle inequalities has only integral vertices. But this is not the case as already observed in [164].

Figure 6.2 shows a fractional vertex for P_C^6. Dotted edges represent pairs of antiparallel arcs the corresponding variables taking the value $\frac{1}{2}$.

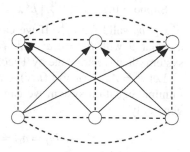

Fig. 6.2 A fractional vertex of P_C^6

The study of this example gives rise to a first class of further inequalities.

Definition 6.1. A digraph $D = (V, F)$ is called a *k-fence* if it has the following properties:

(i) $|V| = 2k, k \geq 3$,
(ii) V can be partitioned into two subsets $U = \{u_1, \ldots u_k\}$ and $L = \{l_1, \ldots, l_k\}$ such that

$$F = \bigcup_{i=1}^{k} \Big(\{(u_i, l_i)\} \cup \{(l_i, u_j) \mid j \in \{1, \ldots, k\}, \ j \neq i\} \Big).$$

Every k-fence $D = (V, F)$ defines the *k-fence inequality* $x(F) \leq k^2 - k + 1$ which is valid for $P_{LO}^n, n \geq 2k$. Figure 6.3 shows a 3-fence.

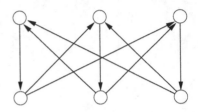

Fig. 6.3 A 3-fence

Note that the fractional vertex shown in Fig. 6.2 violates the inequality associated with 3-fences by $\frac{1}{2}$.

Theorem 6.9. *Let $D = (V, F)$ be a k-fence contained in $D_n, n \geq 2k$. Then the k-fence inequality $x(F) \leq k^2 - k + 1$ defines a facet of P_{LO}^n.*

Proof. To illustrate the technique of how to show, without exhibiting a sufficiently large set of affinely independent vertices, that an inequality is facet defining we give an explicit proof.

Let $D = (V, F)$ be a k-fence, $k \geq 3$. Assume that $U = \{1, 2, \ldots, k\}$ is the set of its upper nodes and $L = \{k+1, k+2, \ldots, 2k\}$ is the set of its lower nodes. We call the arcs $(i, k+i)$, $i = 1, \ldots, k$, *pales* and the other arcs *pickets*. Due to Lemma 6.5 it suffices to show that $x(F) \leq k^2 - k + 1$ is facet defining for P_{LO}^{2k}. Denote this k-fence inequality by $a^T x \leq a_0$.

Suppose that $\{x \in P_{LO}^{2k} \mid b^T x = b_0\} \supseteq \{x \in P_{LO}^{2k} \mid a^T x = a_0\}$ for some inequality $b^T x \leq b_0$ valid for P_{LO}^{2k}. If we can show that there exists a vector $\lambda \in \mathbf{R}^{\binom{2k}{2}}$ and a scalar μ with $b^T = \mu a^T + \lambda^T H$ (where $Hx = \mathbf{1}$ denotes the minimal equation system) then we are done by Theorem 6.1.

Let $\overline{F} = A_n \setminus \{(u, v) \mid (u, v) \in F \text{ or } (v, u) \in F\}$. Because of the structure of the minimal equation system we can make the following assumptions without losing generality

$$b_{ij} = a_{ij} = 1, \text{ for } (i, j) \in F,$$
$$b_{ij} = a_{ij} = 0, \text{ for } (i, j) \in \overline{F} \text{ and } i < j.$$

By reversing $k-1$ pales of D or $k-2$ pales and one picket having no endnode with any of these $k-2$ pales in common we obtain an acyclic digraph which induces a partial ordering of the nodes of D. Every extension of such a partial ordering to a linear ordering gives an acyclic tournament T the incidence vector of which satisfies $a^T \chi^T = a_0$. Let u, v be two nodes in U (or V). If we can find a partial ordering (by reversing exactly $k-1$ arcs of D) which implies neither $u \prec v$ nor $v \prec u$ then there exist linear extensions $T_1 = \langle \alpha, u, v, \beta \rangle$ and $T_2 = \langle \alpha, v, u, \beta \rangle$ satisfying $a^T \chi^{T_1} = a^T \chi^{T_2} = a_0$ and therefore $b^T \chi^{T_1} = b^T \chi^{T_2} = b_0$. From this we get

$$0 = b^T \chi^{T_1} - b^T \chi^{T_2} = b_{uv} - b_{vu}.$$

We first show that $b_{21} = b_{12} = 0$. By reversing all pales except for pale $(k, 2k)$ we obtain a partial ordering of the nodes of D which leaves nodes 1 and 2 incomparable (both nodes are sinks in the corresponding digraph) and hence by the above argument $b_{12} = b_{21} = 0$. Repeating this construction for all pairs of nodes in U, resp. in V yields $b_{ij} = 0$ for all $(i, j) \in \overline{F}$.

We now show that there is some scalar $\xi \in \mathbf{R}$ such that $b_{ij} = \xi$ for all arcs (i, j) with $(j, i) \in F$. W.l.o.g. we assume that $b_{k+1,1} = \xi$

The reversal of all pales except for pale $(2, k+2)$ gives a partial ordering which can be extended to the linear ordering

$$T_1 = \langle k+3, k+4, \ldots, 2k, k+1, 2, k+2, 1, 3, 4, \ldots, k \rangle.$$

The reversal of the picket $(k+2, 1)$ and all pales except for the two pales $(1, k+1)$ and $(2, k+2)$ induces a partial ordering which can be extended to the linear ordering

$$T_2 = \langle k+3, k+4, \ldots, 2k, 1, k+1, 2, k+2, 3, 4, \ldots, k \rangle.$$

From this we get

$$\begin{aligned}
0 &= b^T \chi^{T_1} - b^T \chi^{T_2} \\
&= b_{k+1,1} + b_{21} + b_{k+2,1} - b_{1,k+1} - b_{12} - b_{1,k+2} \\
&= \xi + 0 + 1 - 1 - 0 - b_{1,k+2}
\end{aligned}$$

and hence $b_{1,k+2} = \xi$.

Using similar constructions we eventually obtain $b_{ij} = \xi$ for all arcs antiparallel to the arcs of the k-fence.

Defining $\lambda \in \mathbf{R}^{\binom{2k}{2}}$ with components $\lambda_{\{i,j\}}$ for $i < j$ by

$$\lambda_{\{i,j\}} = \begin{cases} 0, & \text{if } (i, j) \in \overline{F}, \\ \xi, & \text{otherwise}, \end{cases}$$

and setting $\mu = 1 - \xi > 0$, we get $b^T = \mu a^T + \lambda^T H$. This finishes the proof. $\qquad \square$

According to Theorem 6.4 different k-fences induce different facets of P_{LO}^n. Hence the number of facets of P_{LO}^n, $n \geq 6$, which are induced by k-fences is

$$\sum_{k=3}^{\lfloor \frac{n}{2} \rfloor} \left[\binom{n}{2k} \binom{2k}{k} k! \right] = \sum_{k=3}^{\lfloor \frac{n}{2} \rfloor} \frac{n!}{(n-2k)! k!}.$$

For $k > 3$ the k-fences can be considered as a generalization of the 3-fence. Looking at the 3-fence from a different point of view, namely by focusing on the structure of its dicycles, leads to another generalization and the rich class of Möbius ladders. Let $C_1, \ldots, C_k$ be different dicycles in $D_n = (V_n, A_n)$ such that

 (i) C_i and C_{i+1}, for $i \in \{1, \ldots, k-1\}$, have exactly one arc in common. This arc is called e_i. C_k and C_1 have exactly the arc e_k in common.
 (ii) C_i and C_j, for $j \notin \{i-1, i+1\}$ (resp. $j \notin \{k, 2\}$, if $i = 1$), have no common arc.

We define the digraph $D = (V, A)$ by setting $V = \cup_{i=1}^{k} V(C_i)$ and $A = \cup_{i=1}^{k} C_i$ and say that D is *generated* by the dicycles $C_1, \ldots, C_k$. Conversely, whenever a digraph $D = (V, A)$ is said to be generated by k dicycles, we implicitly assume that $V = \cup V(C_i)$, $A = \cup C_i$ and that the common arcs are denoted by e_i as in (i).

A dicycle C_j is called *right-adjacent* to a dicycle C_i if C_j and C_i have some node v in common and if all dicycles C_l contain this node, $l \in \{j, j+1, \ldots, i\}$, for $j < i$, and $l \in \{j, j+1, \ldots, k, 1, 2, \ldots, i\}$, for $j > i$. A dicycle C_j is called *left-adjacent* to a dicycle C_i if C_j and C_i have some node v in common and if all dicycles C_l contain this node, $l \in \{i, i+1, \ldots, j\}$, for $j > i$, resp. $l \in \{i, i+1, \ldots, k, 1, 2, \ldots, j-1, j\}$, for $j < i$. If C_j is both left- and right-adjacent to C_i then all dicycles have a common node.

Figure 6.4 gives a different drawing of the 3-fence showing that it can also be considered a Möbius ladder consisting of three 4-dicycles. The figure also illustrates why the name Möbius ladder was chosen.

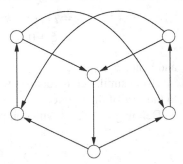

Fig. 6.4 A 3-fence drawn as
Möbius ladder

Not all digraphs of Möbius ladder structure yield facets. They have to be generated by an odd number of dicycles and these dicycles have to be short.

Definition 6.2. Let $D = (V, M)$ be a subdigraph of D_n which is generated by the k dicycles $C_1, \ldots, C_k$, i.e., $V = \cup V(C_i), M = \cup C_i$, and which satisfies the following properties:

(i) $k \geq 3$ and k is odd.
(ii) The length of C_i is three or four, $i = 1, \ldots, k$.
(iii) The degree of each node $u \in V(M)$ is at least three.
(iv) If two dicycles C_i and $C_j, 2 \leq i + 1 < j \leq k$, have a node, say v, in common then C_j is either left-adjacent or right-adjacent to C_i but not both.
(v) Given any dicycle $C_j, j \in \{1, \ldots, k\}$, set $J = \{1, \ldots, k\} \cap (\{j - 2, j - 4, \ldots\} \cup \{j + 1, j + 3, \ldots\})$. Then the set $M \setminus \{e_i \mid i \in J\}$ contains exactly one dicycle, namely C_j.

Then D is called a *Möbius ladder*.

Theorem 6.10. *Let $D = (V, M)$ be a Möbius ladder in D_n generated by the k dicycles $C_1, C_2, \ldots, C_k$. Then the Möbius ladder inequality*

$$x(M) \leq |M| - \frac{k+1}{2}$$

defines a facet of P_{LO}^n for $n \geq |V|$.

Möbius ladder inequalities are mod-2 inequalities w.r.t. to trivial and dicycle inequalities. Namely, let $D = (V, M)$ be generated by $C_1, C_2, \ldots, C_k$ and let $F = \{e_1, \ldots, e_k\}$ be the set of common arcs of adjacent dicycles. Then the addition of the inequalities

$$x(C_1) \leq |C_1| - 1$$
$$\vdots$$
$$x(C_k) \leq |C_k| - 1$$
$$\sum_{e \in F} x_e \leq |M| - |F|$$

gives $2x(M) \leq 2|M| - k$ and hence, after dividing by 2 and rounding down the right hand side, the Möbius ladder inequality $x(M) \leq |M| - \frac{k+1}{2}$.

There are further classes of facet-defining and valid inequalities which we will not describe here:

- *diagonal inequalites* [65],
- Z_m-*inequalities* [143],
- *t-reinforced k-fences* [110],
- *augmented k-fences* [111],
- *Paley inequalities* [75],
- *facets from stability-critical graphs* [100, 47],
- *generalizations of Möbius ladders* [58],
- *new facets by rotations* [18].

Concerning separation, the situation is not too fortunate. The separation problem for k-fences is already NP-hard. The only inequalities so far which can be separated in a systematic way (and in polynomial time) are certain Möbius ladder inequalities as they are mod-2 inequalities [27].

We will discuss how to obtain further cutting planes in the subsequent sections.

6.3 Computation of Complete Descriptions

In addition to identifying classes of facet defining inequalities, there has always been interest in deriving complete linear descriptions for polytopes associated with small instances of combinatorial optimization problems. Though real problems are usually large scale, it is worthwhile to put investigations into small polytopes. Facets derived for small polytopes can give hints for generalizations to facets for larger polytopes and studying small polytopes can also yield information on the relative importance of the different classes of facets. Furthermore, in particular in the case of the LOP (because of the trivial lifting property), facet-defining inequalities for small instances are also facet-defining for large instances and can thus be used in computations.

In general, a $\mathscr{V}$-polyhedron $\mathrm{conv}(V) = \mathrm{conv}(\{v_1 \ldots, v_m\})$ can be transformed to an $\mathscr{H}$-polyhedron according to the following reformulation.

$$\mathrm{conv}(V) = \{x \mid \text{there exists } y \text{ such that } Vy = x, \mathbf{1}^T y = 1, y \geq 0\}$$

$$= \left\{ x \,\middle|\, \text{there exist } y \text{ such that } \begin{pmatrix} V & -I \\ -V & I \\ \mathbf{1}^T & 0 \\ -\mathbf{1}^T & 0 \\ -I & 0 \end{pmatrix} \begin{pmatrix} y \\ x \end{pmatrix} \leq \begin{pmatrix} 0 \\ 0 \\ 1 \\ -1 \\ 0 \end{pmatrix} \right\}$$

$$= \left\{ x \,\middle|\, \text{there exists } y \text{ such that } \begin{pmatrix} y \\ x \end{pmatrix} \in P(D,d) \right\},$$

where D and d are defined by the second-to-last row.

By projecting $P(D,d)$ onto the subspace $y = 0$ (by elimination of the y-variables) we obtain a polyhedron $P(A,b)$ with the property

$$x \in P(A,b) \iff \text{there exists } y \text{ such that } \begin{pmatrix} y \\ x \end{pmatrix} \in P(D,d)$$

$$\iff x \in \mathrm{conv}(V).$$

As a first step one should determine the minimal equation system for $\mathrm{conv}(V)$ by performing Gaussian elimination in the system

$$\begin{pmatrix} V & -I \\ \mathbf{1}^T & 0 \end{pmatrix} \begin{pmatrix} y \\ x \end{pmatrix} = \begin{pmatrix} 0 \\ 1 \end{pmatrix}.$$

The system is brought into an equivalent form

$$\begin{pmatrix} I & * & * \\ 0 & 0 & D' \end{pmatrix} \begin{pmatrix} y \\ x \end{pmatrix} = \begin{pmatrix} * \\ d' \end{pmatrix}$$

with D' as large as possible. The system $D'x = d'$ is an equation system for $\operatorname{conv}(V)$. Possible redundant equations still contained in it can be removed to obtain a minimal system.

$P(A, b)$ can be obtained algorithmically by using Fourier-Motzkin elimination. The following algorithm eliminates the j-th variable from the system $Dx \leq d$ and yields a system $Ax \leq b$ which is defined in the remaining variables only and has feasible solutions if and only if $Dx \leq d$ is feasible.

FourierMotzkin(D, d, j)

(1) Let $k = 0$ and partition the rows $M = \{1, 2, \ldots, m\}$ of D into
$N = \{i \in M \mid d_{ij} < 0\}$,
$P = \{i \in M \mid d_{ij} > 0\}$,
$Z = \{i \in M \mid d_{ij} = 0\}$.
(2) For all $i \in Z$, set $k = k + 1$ and
$A_{k.} = D_{i.}$ and $b_k = d_i$.
(3) For all $(s, t) \in (N \times P)$ set $k = k + 1$ and
$A_{k.} = d_{tj}D_{s.} - d_{sj}D_{t.}$ and $b_k = d_{tj}d_s - d_{sj}d_t$.
(4) The resulting system is $Ax \leq b$.

To turn this principle approach into an effective algorithm further ingredients have to be added, e.g. for avoiding redundancy and finding good elimination orders. All details can be found in [39]. Here we report on the most important findings for the linear ordering polytope.

The complete linear description of P_{LO}^6 consists of the 15 equations forming the minimal equation system, 30 trivial inequalities, 40 3-dicycle inequalities, 120 3-fence inequalities and 360 inequalites from each of the two types of Möbius ladders M_1 (defined on four 3-dicycles and one 4-dicycle) and M_2 (the reversal of M_1).

Figure 6.6 displays two facet defining inequalities of P_{LO}^7. All solid arcs shown have coefficient 1 in the inequality, the dotted arc has coefficient 2. The right hand side of the inequality on the left side is 13 and the other inequality has right hand side 14.

The complete linear description of P_{LO}^7 given in [144] consists of 87,472 facets. To visualize how many facets are structurally different we define equivalence classes. Two facet defining inequalities are said to belong to the same σ-*class* if one can be obtained from the other just by renaming the nodes. The vertices satisfying a facet defining inequality with equation are called *roots* of the facet.

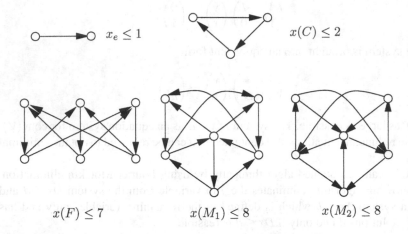

Fig. 6.5 Facet defining inequalities of P_{LO}^6

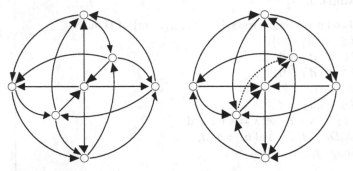

Fig. 6.6 Facet defining inequalities for P_{LO}^7

The facets of P_{LO}^7 can be partitioned into 27 σ-classes which are displayed in Table 6.1. Eight of these classes are obtained from other classes just by arc reversal. They are marked with a "*". The table also gives the number of vertices on each facet and the number of nonequivalent facets in each class.

Table 6.1 substantiates the importance of trivial (F1) and 3-dicycle (F2) inequalities. Every such inequality defines a facet with 2520 roots, i.e., a facet containing half of the vertices of P_{LO}^7. A 3-fence facet (F3) contains only 126 vertices and there are facets containing only a few roots more than required for the facet dimension.

A further partition of facets of P_{LO}^n into equivalence classes is possible. Let $P = \text{conv}(V)$ be a polytope in $\mathbf{R}^d$. An affine mapping ψ from $\mathbf{R}^d$ to itself is called a *rotation mapping for* P if $\psi(V) = V$. A rotation mapping transforms a facet F of P to a facet $\psi(F)$.

Table 6.1 The polyhedral structure of P_{LO}^7

Class	#vertices on facet	#different facets
F1	2,520	42
F2	2,520	70
F3	126	840
F4, F4*	126	5,040
F5, F5*	67	10,080
F6	44	5,040
F7, F7*	104	5,040
F8, F8*	67	10,080
F9, F9*	67	10,080
F10, F10*	44	5,040
F11	67	5,040
F12	104	5,040
F13	67	5,040
F14	126	840
F15, F15*	104	5,040
F16	104	5,040
F17, F17*	28	2,520
F18	28	2,520
F19	28	5,040

Arc reversal is one such rotation mapping. A second one was introduced in [18]. For fixed $r \in \{1,\ldots,n\}$ the rotation mapping ψ is defined by

$$\psi(x)_{rj} = x_{jr}, \text{ for all } 1 \le j \le n, j \ne r,$$

$$\psi(x)_{jr} = x_{rj}, \text{ for all } 1 \le j \le n, j \ne r,$$

$$\psi(x)_{ij} = x_{ij} + x_{jr} + x_{ri} - 1, \text{ for all } 1 \le i, j \le n, i \ne r, j \ne r.$$

It was shown in [18] that if $a^T x \le \alpha$ defines the facet F of P_{LO}^n then

$$\sum_{i=1, i \ne r}^{n} \sum_{j=1, j \ne r}^{n} a_{ij} \psi(x)_{ij}$$

$$= \sum_{i=1, i \ne r}^{n} \left(\sum_{j=1, j \ne r}^{n} a_{ij}(x_{ij} + x_{jr} - x_{ir}) + a_{ir}x_{ri} + a_{ri}x_{ir} \right)$$

$$\le \alpha$$

defines the facet $\psi(F)$ of P_{LO}^n. E.g. this mapping transforms a 3-dicycle inequality to a trivial inequality and vice versa. We say that two facets belong to the same P_{LO}^n-*class* if they can be converted to each other either by renaming the nodes or by applying a rotation mapping.

Using a parallel computer and the adjacency approach described in [39] we could compute at least a lower bound on the number of facets of P_{LO}^8. We could not prove that this lower bound gives the exact number of facets, because in 11 out of 12,231 cases we could not compute the adjacent facets. These cases are the facets with the maximum number of roots (among them the facets defined by trivial, 3-dicycle, 3-fence and 4-fence inequalities). At least 67.5% of the facet σ-classes were first discovered by our computations. This is a very conservative estimation, based on the observation that no facets of P_{LO}^8 with coefficients larger than 2 were known before, while for that percentage of facet σ-classes the minimal coefficient is 3.

Table 6.2 summarizes the current state of knowledge about linear ordering polytopes.

Table 6.2 Facet structure of P_{LO}^n

n	#vertices	#different facets	#σ-classes	P_{LO}^n-classes
3	6	8	2	1
4	24	20	2	1
5	120	40	2	1
6	720	910	5	2
7	5,040	87,472	27	6
8	40,320	$\geq 488,602,996$	$\geq 12,231$	$\geq 1,390$

It is impressive how huge the number of facets of combinatorial polytopes is even for small instances. And, moreover, none of them is superfluous and they are also present in larger instances. We will therefore below turn to the question whether we can make use of this wealth of inequalities in practical computations.

6.4 Differences between Facets

It is a natural question which classes of facets are most or least useful in cutting plane algorithms. For reasons of efficiency, the choice of the inequalities to use in a cutting plane algorithm is typically dictated by whether efficient exact or heuristic algorithms are known for the corresponding separation problem. Up to now no ultimate measure of the quality of a valid inequality with respect to its application in a branch-and-cut algorithm could be established. From our computations of descriptions of small polytopes we now have 12231 different classes of (necessary!) facet defining inequalities available for all linear ordering problems on at least 8 nodes. Even on parallel hardware it does not seem to make sense to call a separation procedure for every class. In the following we therefore want to exhibit differences between the facets in order to possibly obtain insight into their usefulness for branch-and-cut algorithms.

Our goal is to find out if there is a measure such that a priority rule can be given for the application of the facets of a small problem instance relaxation in a branch-and-cut algorithm.

In [74, 75] the notion of *strength of a relaxation* was introduced. The strength of a relaxation is meant to be a measure of how well a relaxation approximates a poly-hedron in comparison to another weaker relaxation. The strength is only defined for certain types of combinatorial polyhedra, namely polyhedra of blocking type [74] and of anti-blocking type [75]. In the case of the LOP the anti-blocking type is of interest.

Definition 6.3. Given polytopes P and Q of anti-blocking type, P is said to be an α-relaxation of Q for some $\alpha \geq 1$ if $Q \supseteq \frac{1}{\alpha}P = \{\frac{x}{\alpha} \mid x \in P\}$. The strength $\text{str}(P, Q)$ of Q with respect to P is the minimum value of α such that P is an α-relaxation of Q.

Notice that $\text{str}(P, Q) \geq 1$ and $\text{str}(P, Q) = 1$ if and only if $P = Q$. In general, $\text{str}(P, Q)$ could be infinite.

While P_{LO}^n is not of blocking or anti-blocking type we can apply this concept because of its close relation to the acyclic subdigraph polytope P_{AC}^n. The *acyclic subdigraph polytope*

$$P_{AC}^n = \text{conv}\{\chi^B \in \{0, 1\}^{n(n-1)} \mid B \text{ is an acyclic arc set in } A_n\}$$

is of anti-blocking type.

Because $P_{LO}^n = \{x \in P_{AC}^n \mid x_{ij} + x_{ji} = 1, 1 \leq i < j \leq n\}$, the linear ordering poly-tope P_{LO}^n is a face of P_{AC}^n. For any nonnegative objective function, an optimal solu-tion of the linear ordering problem is an optimal solution of the acyclic subdigraph problem.

Furthermore, a facet defining inequality for P_{LO}^n in normal form is valid, resp. facet defining for P_{AC}^n. Hence, if we restrict ourselves to nonnegative objective func-tions (which we can do w.l.o.g. for the LOP), we can interpret a cutting plane algo-rithm for solving the *linear ordering problem* as a cutting plane algorithm for the *acyclic subdigraph problem* using exclusively facet defining inequalities for P_{LO}^n. This observation justifies the discussion of strength of relaxations given by inequal-ities for P_{LO}^n.

The *strength of an inequality* with respect to a polytope P is the strength of the relaxation obtained by adding the inequality. We will compute the strength of classes of inequalities with respect to the following polytopes.

Trivial relaxation:

$$P_T^n = \{x \mid x_{ij} + x_{ji} = 1, \ 1 \leq i, j \leq n,$$
$$x_{ij} \geq 0, \ 1 \leq i, j \leq n\}.$$

Dicycle relaxation:

$$P_C^n = \{x \mid x_{ij} + x_{ji} = 1, \ 1 \le i, j \le n,$$
$$x_{ij} \ge 0, \ 1 \le i, j \le n$$
$$x(C) \le 2, \ \text{for all dicycles } C \text{ of length } 3\}.$$

Obviously, $P_C^n = P_T^n \cap \{x \mid x(C) \le 2, \text{ for all dicycles } C \text{ length } 3\}$.

Let $f^T x \le f_0$ be a facet defining inequality for P_{LO}^n in normal form. Following [75] we define the *trivial strength* of this inequality as

$$s_T(f) = \frac{\max\{f^T x \mid x \in P_T^n\}}{\max\{f^T x \mid x \in P_{LO}^n\}} = \frac{\max\{f^T x \mid x \in P_T^n\}}{f_0}$$

and the *dicycle strength* of f as

$$s_C(f) = \frac{\max\{f^T x \mid x \in P_C^n\}}{\max\{f^T x \mid x \in P_{LO}^n\}} = \frac{\max\{f^T x \mid x \in P_C^n\}}{f_0}.$$

Note that $\max\{f^T x \mid x \in P_T^n\} = \mathbf{1}^T f$ since $f^T x \le f_0$ is in normal form.

Figure 6.7 shows the trivial strengths of the σ-classes of facets of P_{LO}^8 (except for the trivial inequalities).

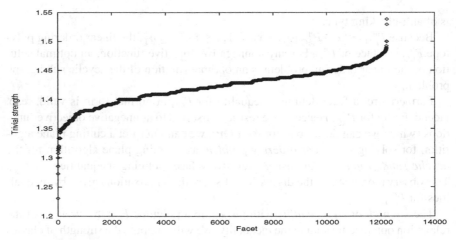

Fig. 6.7 Trivial strength of σ-classes of facets of P_{LO}^8

Figure 6.8 displays the only two facets of P_{LO}^8 with $s_T > 1.5$. The right hand sides of these facets are 17 and 13, the trivial strengths being $s_T = 1.52941$ and $s_T = 1.53846$, respectively. The 3-dicycle inequality has trivial strength 1.5.

Figure 6.9 displays the dicycle strength of the σ-classes of facets (except for the trivial and 3-dicycle inequalities). The values of $\max\{f^T x \mid x \in P_C^n\}$ were computed using the branch-and-cut algorithm for the linear ordering problem described earlier.

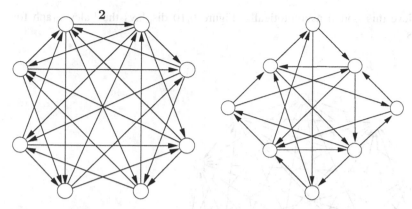

Fig. 6.8 Facet defining inequalities with $s_T = 1.52941$ and $s_T = 1.53846$

The 4-fence inequality is the facet with maximum dicycle strength $s_C = 1.07692$ among all facets of P^8_{LO}.

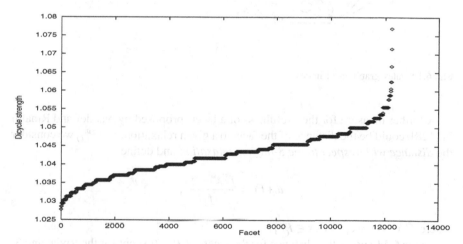

Fig. 6.9 Dicycle strength of σ-classes of facets of P^8_{LO}

As shown by Goemans and Hall [75], the trivial strength of P^n_{AC} is $2 - o(1)$, where $o(1)$ is nonnegative and tends to 0 as the number of nodes n tends to infinity. They prove that the highest trivial strength of the known inequalities of P^n_{LO} is attained asymptotically for the augmented k-fence inequality with a value of only 1.52777 and they present new valid inequalities, called Paley inequalities, which they prove to be facet defining for $n = 11$ and $n = 19$ with trivial strengths 1.57143 and 1.59813, respectively. They conclude that the strongest facets of the acyclic subdigraph polytope are unknown. Concerning the dicycle strength they show that the value for P^n_{AC} must be at least $\frac{4}{3}$ since the Paley inequalities achieve this bound asymptotically. Figure 6.10 displays the Paley graph for 11 nodes.

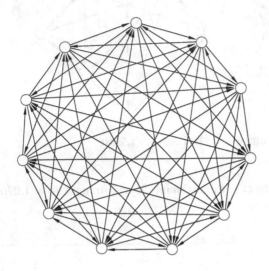

Fig. 6.10 Paley graph on 11 nodes

A further measure for the usefulness of a facet (proposed by Naddef and Rinaldi in [128]) could be the distance of the facet to a given relaxation. For P_{LO}^n we consider the *distance with respect to the 3-dicycle relaxation* and define

$$d_c(f) = \frac{f^T x^* - f_0}{|f|},$$

where $f^T x^* = \max\{f^T x \mid x \in P_C^n\}$.

Figure 6.11 shows this distance for the facets of P_{LO}^8 (except for the trivial and 3-dicycle inequalities). The strongest facet in this sense is the 4-fence inequality with dicycle distance $d_c = 0.25$.

A further indication of the usefulness of a facet could be its volume. Volume computation is very difficult and we refer to [39] where volume computation methods are described and where it is shown that there is a correlation between the volume of facets and their number of roots.

If one examines correlations between these measures then it turns out that dicycle strength and dicycle distance are correlated whereas there seems to be a negative correlation between trivial and dicycle strength, i.e., a facet with large dicycle strength has small trivial strength and vice versa. There seems to be a weak correla-

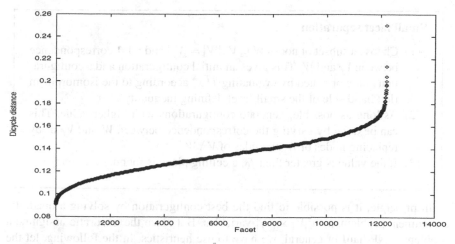

Fig. 6.11 Distance of σ-classes of facets of P_{LO}^8 relative to P_C^8

tion between number of roots and trivial strength but no correlation between number of roots and dicycle strength or distance.

6.5 Separation of Small Facets

We now describe a possibility for employing the "small" facets in practical computations. We do not attempt to generalize promising classes because the separation problem for well-structured inequalities like fences and Möbius ladders is already hard. Of course, the study of generalizations is an interesting topic for theoretical research. We rather develop a general procedure which has the chance to find violated inequalities of any facet class.

Let $f^T x \leq f_0$ be a small facet defining inequality for the LOP (in normal form) with support graph $D_f = (V_f, A_f)$. If x^* is the current fractional solution, we have to identify a subgraph $D' = (V', A')$ of D_n with $|V'| = |V_f|$ and an isomorphism between V' and V_f such that $f^T x^* > f_0$ on this subgraph.

We use the following method for identifying the subset V' and the isomorphism.

Small facet separation

(1) Choose a subset of nodes $W \subseteq V$, $|W| = |V_f|$ and a 1-1 correspondence between V_f and W. This gives an initial configuration and a configuration value obtained by evaluating $f^T x^*$ according to the isomorphism (left hand side of the small facet defining inequality).

(2) As long as possible, generate configurations with higher value. This can be done by varying the correspondence between W and V_f or by replacing nodes of W by nodes of $V \setminus W$.

(3) If the value is greater than f_0, a cutting plane is found.

In principle, it is possible to find the best configuration by solving a quadratic assignment problem with $|V| \times |V_f|$ variables. But since the quadratic assignment problem is NP-hard in general, we have to use heuristics. In the following, let the current configuration be represented by a mapping $\sigma : V \to V_f \cup \{0\}$ with $\sigma(w) = u$, if $w \in W$ and u is the node of V_f associated with w in the 1-1 correspondence, and $\sigma(w) = 0$, if $w \in V \setminus W$.

Figure 6.12 illustrates the principle problem and how to proceed with local modifications. Here a violated facet defining inequality from a facet class with a support graph on 7 nodes is searched for.

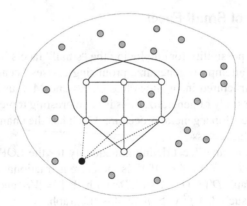

Fig. 6.12 Separation of small facets

We can use virtually any of the heuristic or meta-heuristic approaches described in this book. For our computational experiments we have chosen several methods. Here we briefly describe a deterministic improvement and a simulated annealing heuristic. A more elaborated GRASP algorithm is given in [39].

Deterministic Improvement

(1) For every pair of nodes $u, v \in V$ perform the following:

   (1.1) Define a new configuration σ' by setting $\sigma'(u) = \sigma(v)$, $\sigma'(v) = \sigma(u)$, and $\sigma'(w) = \sigma(w)$ for $w \in V \setminus \{u, v\}$.
   (1.2) If the new configuration has higher value, set $\sigma = \sigma'$.

(2) If the current configuration could be improved for at least one pair, then repeat step (1).
(3) For every triple of nodes $u, v, w \in W$ perform the following:

   (3.1) Consider all possible new configurations that can be obtained by exchanging the assignments of u, v and w.
   (3.2) If one of these new configurations, say σ', has higher value than the current configuration, set $\sigma = \sigma'$.

(4) If the current configuration could be improved for at least one triple, then repeat step (1).

This deterministic heuristic follows the principle of local search and can be characterized as a combination of 2-node and 3-node exchanges. Obviously, the modification capabilities of this heuristic are limited, but based on experience from many local search algorithms one can expect that reasonable configurations are determined.

In our separation approach we do not exploit special structural information about small facets to find good configurations. For such unstructured searches, usually also meta-heuristic algorithms give good results. The second approach is a simulated annealing scheme.

Simulated Annealing

(1) Choose an initial parameter ϑ, a repetition factor r, and a stopping parameter ε.
(2) Perform the following as long as $\vartheta > \varepsilon$:

   (2.1) Repeat the following steps r times
      (2.1.1) Let σ be the current configuration. Choose two nodes $u, v \in V$ at random, and define a new configuration σ' by setting $\sigma'(u) = \sigma(v)$, $\sigma'(v) = \sigma(u)$, and $\sigma'(w) = \sigma(w)$ for $w \in V \setminus \{u, v\}$. Let Δ be the difference of the new configuration value and the old one.
      (2.1.2) If $\Delta > 0$ set $\sigma = \sigma'$.
      (2.1.3) If $\Delta \leq 0$, compute a random number p, $0 \leq p \leq 1$ and set $\sigma = \sigma'$ if $p \leq \frac{e^{\Delta}}{\vartheta}$.

(3) Update ϑ and r.

To give an impression of how many violated inequalites one can expect from small facet separation we list in Tables 6.3 and 6.4 the number of inequalities found for the facet classes of P_{LO}^7.

Table 6.3 Number of small facets of classes F3–F11 found

Problem	F3	F4	F5	F6	F7	F8	F9	F10	F11
econ59	26	163	11	1	1	–	105	3	–
econ64	1	47	1	–	1	–	9	3	–
econ71	1	8	–	–	–	–	35	–	–
econ72	19	34	2	2	5	3	66	8	–
econ76	4	147	1	–	1	–	17	3	–
econ77	–	42	9	–	25	17	49	10	–
randB	195	1435	41	11	103	82	134	28	50
randC	130	1035	26	2	55	37	54	11	30

In these experiments problem instances were selected which could not be solved at the root node with 3-dicycle inequalities only. Small facet separation was invoked when no more 3-dicycle inequalities were violated. Now, all problems could be solved at the root node.

Table 6.4 Number of small facets of classes F12–F19 found

Problem	F12	F13	F14	F15	F16	F17	F18	F19
econ59	1	6	108	67	11	102	15	–
econ64	1	–	18	6	3	5	1	–
econ71	1	3	26	24	1	13	–	–
econ72	1	1	79	34	19	55	6	–
econ76	–	–	28	18	–	9	2	–
econ77	–	4	141	24	–	74	17	–
randB	75	23	107	134	70	1	4	–
randC	36	17	36	63	25	2	3	–

For the linear ordering problem we make use of the fact that the facets of small polytopes are globally facet defining even when lifted trivially. Fractional LP solutions for the linear ordering problem are dense, since we are working on the complete directed graph and since a solution vector contains $\binom{n}{2}$ arcs with value 1. Therefore, we could not develop a reasonable shrinking procedure and perform small facet separation in the complete digraph with arc weights given by the fractional solution.

6.6 Computational Experiments with Small Facets

Extensive computational experiments have been carried out to find the best strategy for employing facets from small relaxation. They are documented in depth in [39, 40] and we only cite the major insights here.

The linear ordering problems were always solved as follows. In a first phase, only 3-dicycle inequalities were generated. If not enough 3-dicycle cuts are found, then the heuristics for small facet generation are invoked. Because of the huge number of facet classes various approaches have been tested. All computations were carried out on a PC cluster where one processor was the master processor handling the core of the branch-and-cut and the other processors solved the separation problem for small facets. The facet classes taken into account were partitioned on the available processors.

6.6.1 Comparison of Heuristics

It turned out that simulated annealing and GRASP were more effective that the deterministic approach. Annealing and GRASP were of about the same quality with respect to finding inequalties. Because of the easier tuning of parameters, eventually GRASP was preferred.

6.6.2 Cutting Plane Selection

Usually, at each LP phase in the branch-and-cut algorithm many violated inequalities were generated. This was in particular the case because many parallel processors were active. It is well-known that it is not advisable to add all violated inequalities because the linear programs become more difficult. The following strategies were tested.

S1 Select some cuts at random.
S2 Select cuts with priority depending on the amount of violation $\delta = f^T x^* - f_0$ (higher violation preferred). This strategy is called *distance*.
S3 Select cuts with priority depending on the distance d between x^* and the hyperplane $f^T x = f_0$, i.e., $d = \frac{f^T x^* - f_0}{\|f\|} = \frac{\delta}{\|f\|}$.
S4 If $f = kc$ with $k > 0$, then $f^T x \leq f_0$ is obviously the "best" inequality that can be added to the linear program, since f_0 is an upper bound of the objective function which is attained for the roots of $f^T x \leq f_0$. Therefore, an idea is to prefer cutting planes $f^T x \leq f_0$ with f as parallel as possible to the objective function c, i.e., those which maximize $\frac{c^T f}{\|f\|}$. Moreover, it might be assumed that a hyperplane $f^T x = f_0$ which is nearly parallel to the direction of the objective

function can bound more symmetrically in all directions when reoptimizing. Since the angle ϕ between c and f satisfies $\cos\phi = \frac{c^T f}{\|c\|\|f\|}$, we call this strategy *angle*.

S5 Add cutting planes such that the "expected" following LP solution y^* minimizes $c^T x$. Since, of course, y^* is not known before reoptimizing, we take $-f$ as approximating the direction of reoptimization. Let $\delta = f^T x^* - f_0$. Then with $y^* = x^* - \lambda f$ and $f^T y^* = f_0$ one obtains $\lambda = \frac{\delta}{f^T f}$ and hence

$$c^T y^* = c^T x^* - \frac{\delta c^T f}{f^T f}.$$

Our experiments suggest using strategy S4. It usually leads to the least number of subproblems and also to the least CPU time. Note that this is also the preferrable cut selection strategy if only the 3-dicycle relaxation is solved.

6.6.3 Number of Classes Taken into Account

Of course, the gap between the LP bound and integer optimum decreases if more cutting planes are available. Figure 6.13 displays the gap closure achieved when using more and more facet classes where in the final case all classes are employed. It is seen that there is almost no effect any more if more than 1000 classes are used.

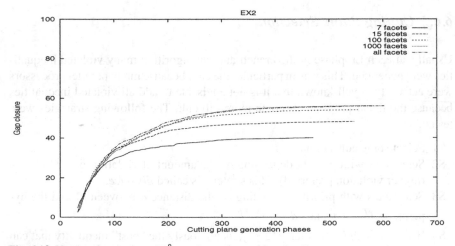

Fig. 6.13 Varying the number of P_{LO}^8 σ-classes for problem (EX2)

6.6.4 Facet Selection

A next point to be addressed is the question which of the 12229 σ-classes of facets of P_{LO}^8 should be used for separation in addition to trivial and 3-dicycle inequalities. We have seen above that they have different properties and it should be found out which of them are important.

In the experiments the 100 best facets relative to the following four criteria were chosen:

  C1 number of roots,
  C2 trivial strength,
  C3 dicycle strength,
  C4 dicycle distance.

Interestingly, it turned out that the number of roots (supposedly correlated to the volume) does not generally outperform the other choices. It seems that the trivial strength is a better indicator for the usefulness of a facet. But it was also observed that this is highly problem dependent.

As the final outcome of all experiments it turned out that, if CPU time in the parallel setup should be minimum, only the 14 facet classes with most roots should be taken into account and the 300 cuts best with respect to criterion "angle" or "distance" should be added in each phase. Despite parallelism, separation of small facets is still very time consuming and only one or two such phases where introduced at every node of the branch-and-cut tree. But, since all cuts found are globally valid, a pool of cuts was kept which could be checked by every node.

In this way the total elapsed time could be reduced by 50% compared with a sequential implementation using 3-dicycles only. Note that, when using 3-dicycles only, there is no reasonable use of parallelization (except for parallel processing of the tree), whereas it makes no sense to use small facet separation in a sequential code. The number of nodes in the branch-and-cut tree was significantly reduced. At the expense of more CPU time the number of nodes could be reduced further, but the above strategy seems to give the best trade-off between less CPU time or less number of nodes in the tree. The separation of small facets proved to be useful, and more research on their optimal employment is needed. However, it seems that, due to their small number of nonzero coefficients, they are not the key to achieving a breakthrough for solving really large problems.

6.7 Local Cuts and Target Cuts

There is a further general possibility for finding violated inequalities which is in particular easily applicable for the LOP. This approach is somehow related to the small facet cut generation as discussed above in that the original problem and the associated fractional solution are projected to a smaller problem, and cuts for the smaller problem are lifted back to the orginal one. The difference is that not "template cuts" (i.e., inequalities which are basically known) are sought, but that general cuts can be derived. The approach for generating *local cuts* was invented and extensively used for solving large traveling salesman problems [6]. We only describe the principal idea.

Assume that the given problem is large and that for the current fractional solution x^* either no cut can be found, or searching for cuts would be much too time consuming because of the size of the problem.

The idea now is to go to a smaller problem instance by some way of "shrinking" the original feasible set P to a smaller set Q, at the same time transforming x^* to a point y^*. The goal is to do the shrinking such that, if x^* is infeasible for P, then also y^* is infeasible for Q. Now it is tried to find an inequality $b^T x \leq b_0$ separating y^* from Q which can basically be accomplished by solving the problem

$$\max b^T y^* - b_0$$
$$b^T y - b_0 \leq 0, \text{ for all } y \in Q.$$

In the case of violation (solutions with objective function value greater than 0 exist), this problem would be unbounded, so some normalization conditions are needed for b. E.g., this could be realized by bounding the components of b, but we do not elaborate on this here.

Because of the trivial lifting property of the linear ordering polytope we can assume that Q is defined by a subset of the nodes of the original problem (i.e., consists of the characteristic vectors of all linear orderings of the nodes of this subset) and that y^* is just x^* restricted to this subset.

The new problem can now be solved iteratively in the following way. We start with some linear orderings $\chi^1, \ldots, \chi^k \in Q$ giving a subset $\bar{Q} \subseteq Q$. Then the linear programming problem (augmented by normalization conditions)

$$\max b^T y^* - b_0$$
$$b^T \chi^i - b_0 \leq 0, \text{ for all } i = 1, \ldots, k$$

is solved. Let z^* be the optimum objective function value with corresponding inequality $b^{*T} x \leq b_0^*$. If $z^* \leq 0$ then no violated inequality can be found for Q. If $z > 0$ then it has to be checked whether the computed inequality is valid for Q or not. In the first steps one can try to find points in Q violating this inequality by running heuristics for finding linear orderings with high value w.r.t. the objective function b^*. If violating linear orderings are found then they can be added to $\bar{Q}$ and the above problem is solved again. If the heuristic does not find a violating ordering then the problem has to be solved exactly by optimizing over Q. This amounts to solving a small LOP. If the optimum objective function value is still positive, then a cutting plane for the small problem has been found, otherwise the shrinking procedure has failed to generate a cut.

If a cut has been found, it has to be lifted to a valid inequality $a^T x \leq a_0$ for the original problem. In the case of the LOP it can be taken as is and a cutting plane for the original problem can be added.

There is a slight addition to this procedure which is helpful in practical computations. Usually the cut is tight just at a single vertex of Q. In order to get a better inequality this cut has to be "tilted" to include further vertices until eventually a facet cut is constructed.

With the so-called *target cuts* the same intention as with local cuts is followed. We want to find cuts for large problems by finding cuts for a problem originating from the true one by some shrinking procedure.

Let P, x^*, Q and y^* be defined as above.

In a first step, all linear orderings $\chi^1, \ldots, \chi^m \in Q$ are enumerated. Let x^0 be an interior point of $\text{conv}(\{\chi^1, \ldots, \chi^m\})$ (e.g., $x^0 = \frac{1}{m}\sum \chi^i$). Note that in our case $\text{conv}(\{\chi^1, \ldots, \chi^m\}) = P_{LO}^k$, if Q is defined by k nodes.

The separation problem can be solved by determining the optimum solutions of the following pair of dual linear programs.

$$\min \sum_{i=1}^{m} \lambda_i$$

$$\sum_{i=1}^{m} \lambda_i (\chi_i - x^0) = x^* - x^0$$

$$\lambda_i \geq 0$$

$$\max (x^* - x^0)^T u$$

$$(\chi_1 - x^0, \ldots, \chi_m - x^0)^T u \leq 1$$

If $\sum \lambda_i^* > 1$ for the pair λ^* and u^* of optimum solutions then the violated inequality $u^{*T}(x - x^0) \leq 1$ is found. This inequality is facet defining for P_{LO}^k.

The main difference between the two approaches is that target cuts generate facets for the subproblem while local cuts generate facets only after tilting. But on the other hand, local cut generation can work with subsets of Q (*delayed column generation*), and a cut can be found without enumerating the set Q. Delayed column generation can also be adopted for target cuts, but this is more complicated (details can be found in [21]).

Chapter 7
Further Aspects

Abstract In this chapter we want to address some issues of interest for the LOP which we have not included in the previous chapters and point to some areas for possible further research. Additionally, in the last section, we consider some extensions of the MDP that include side constraints to model real problems in location and business.

7.1 Approximative Algorithms

It is surprising that, although the LOP is a classical difficult combinatorial optimization problem, not much is known about heuristics with approximation guarantees.

For a nonnegative objective function, a trivial heuristics guarantees an approximation of 50%. Namely, take a random ordering, then either the arc weights of the acyclic tournament induced by this ordering, or the arc weights of the reverse tournament sum to at least half of the total weight, and so $\frac{1}{2}$-approximation is trivial. In [132] it is proved that it is NP-hard to approximate the LOP with a factor better than $\frac{65}{66}$.

As a somewhat contrasting result, it was shown in [23] that the LOP is "asymptotically easy". To be more precise: under certain mild probability assumptions, the ratio between the objective function values of the best and of the worst solution is arbitrarily close to 1 with probability tending to 1 if the problem size goes to infinity. On the other hand, intuitively this result is not surprising. If all entries of the matrix are drawn from a uniform distribution then there is no specific structure and all permuted matrices look similar. And, since one half of the entries are added, the expected difference between values of feasible solutions should be small.

But, as we have seen in the preceding chapters, it is not very difficult to come up with solutions close to optimal for arbitrary instances using elaborate heuristics. This suggests that better approximation results should be possible to narrow the gap $[\frac{1}{2}, \frac{65}{66}]$ which is so far an open problem for possible polynomial time approximation.

© Springer-Verlag GmbH Germany, part of Springer Nature 2022
R. Martí and G. Reinelt, *Exact and Heuristic Methods in Combinatorial Optimization*, Applied Mathematical Sciences 175,
https://doi.org/10.1007/978-3-662-64877-3_7

There are some approximation results on special variants of the problem like
the minimum feedback arc set or the acyclic subdigraph problem. For a survey of
approximation results and a discussion of further interesting aspects and open questions concerning tournaments see [36].

7.2 Integrality Gaps of LP Relaxations

The key to solving the LOP with branch-and-bound or branch-and-cut algorithms
is the strength of the relaxations employed. There has been interesting research on
the so-called *integrality gap* of LP relaxations. For some given nonnegative weight
function c, let $c_{opt}(c)$ and $c_{LP}(c)$ be the corresponding values of the optimum linear
ordering and of the LP relaxation, respectively. Then the integrality gap of the LOP
w.r.t. this relaxation is defined as

$$\sup_{c>0} \frac{c_{LP}(c)}{c_{opt}(c)}.$$

In [132] it is shown that the gap of the standard LP relaxation with 3-dicycle inequalities is arbitrarily close to 2. The proof is based on the existence of a class of
digraphs $D_\varepsilon = (V,A)$ with the property that any acyclic subset of the arcs does not
contain more than approximately $(1+\varepsilon)\frac{|A|}{2}$ arcs.

But, also when further facet-defining inequalities are added, this gap is not improved too much. If all k-fence inequalities were added, the integrality gap would
still be $2-\varepsilon$. The situation could be a little better when small Möbius ladder inequalities (on up to 7 nodes) are added. In this case, the integrality gap can be shown to
be at least $\frac{33}{17} - \varepsilon$, but it could be strictly bounded away from 2.

Based on the approximation results of [132] cited above we can conclude that,
unless $P = NP$, no LP relaxation can have an integrality gap less than $\frac{66}{65}$ if it is
solvable in polynomial time.

The integrality gap of the 3-dicycle relaxation is closely related to the dicycle
strength discussed in the previous chapter. Namely, if the objective function is given
as the left hand side of a facet-defining inequality in normal form, then the integrality
gap is exactly the dicycle strength of this facet.

7.3 Degree of Linearity

As a measure for the "triangularity" of an (n,n)-matrix $C = (c_{ij})$ the *degree of linearity* is defined as

$$\lambda(C) = \frac{\sum\limits_{\sigma(i)<\sigma(j)} c_{ij}}{\sum\limits_{i\neq j} c_{ij}}$$

for an optimum permutation σ. Consider the trivial relaxation of the LOP:

$$\max \sum_{(i,j)\in A_n} c_{ij}x_{ij}$$

$$x_{ij} + x_{ji} = 1, \text{ for } 1 \leq i < j \leq n,$$

$$x_{ij} \in \{0,1\}, \text{ for } 1 \leq i, j \leq n, i \neq j.$$

This relaxation is called the *tournament relaxation* because its feasible solutions are exactly the tournaments in A_n. It can be solved simply by setting, for $i < j$, $x_{ij} = 1$ and $x_{ji} = 0$, if $c_{ij} \geq c_{ji}$, and $x_{ij} = 0$ and $x_{ji} = 1$, otherwise. Therefore, the constraint "$x_{ij} \in \{0,1\}$" can be replaced by "$x_{ij} \geq 0$" and the relaxation can also be viewed as an LP relaxation.

For a problem in normal form, the optimum value is just the sum of all weights. If $c_T(c)$ denotes this optimum value and $c_{\mathrm{opt}}(c)$ the maximum weight of an acyclic tournament, then

$$\frac{c_T(c)}{c_{\mathrm{opt}}(c)} = \frac{1}{\lambda(C)},$$

i.e., the integrality gap of this relaxation is the inverse of the degree of linearity, and thus lies in the interval $[1,2]$. The results on the integrality gap of the 3-dicycle relaxation show that the worst case bound 2 is indeed tight.

In analogy with the 3-dicycles, we now have a connection with the trivial strength of facets. Namely, if the objective function is given as the left hand side of a facet-defining inequality in normal form, then the integrality gap of the tournament relaxation is exactly the trivial strength of this facet.

The integrality gaps of the tournament and of the 3-dicycle relaxation should be correlated. Figure 7.1 displays the corresponding integrality gaps for a set of random problems with $n = 44$.

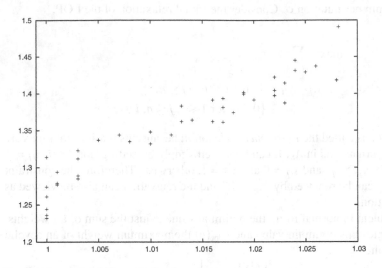

Fig. 7.1 Correlation between 3-dicycle and tournament gap

There is a significant correlation but, e.g., for instances where the 3-dicycle gap is 1, the tournament gap varies between 1.232 and 1.314. Note that the gaps are far from their worst case bound.

We have performed some experiments for learning more about the distribution of the degree of linearity. To this end we have generated random problems in normal form with different densities and sizes of coefficients.

Tables 7.1–7.3, respectively, show minimum, maximum and average degrees of linearity for random problems with densities from 10% to 100% and integral coefficients drawn uniformly from the interval $[0, U]$, where $U = 1, 2, 20, 200, 2000$. For every combination 1000000 instances were solved.

Table 7.1 Miminum degree of linearity ($n = 14$, 1000000 problems)

	10%	20%	30%	40%	50%	60%	70%	80%	90%	100%
1	0.667	0.750	0.744	0.721	0.720	0.702	0.700	0.691	0.683	0.681
2	0.667	0.759	0.750	0.740	0.730	0.720	0.707	0.703	0.689	0.686
20	0.694	0.769	0.772	0.764	0.750	0.731	0.725	0.719	0.710	0.701
200	0.690	0.765	0.777	0.767	0.740	0.734	0.729	0.712	0.705	0.701
2000	0.679	0.765	0.776	0.758	0.750	0.730	0.723	0.720	0.713	0.701

We conducted further experiments with smaller and larger problems. On average, the following tendencies can be observed and are confirmed by the results shown here:

– Problems with smaller entries have lower degree of linearity.

Table 7.2 Maximum degree of linearity ($n = 14$, 1000000 problems)

	10%	20%	30%	40%	50%	60%	70%	80%	90%	100%
1	1.000	1.000	1.000	1.000	1.000	1.000	0.967	0.944	0.923	0.923
2	1.000	1.000	1.000	1.000	1.000	1.000	0.974	0.956	0.929	0.910
20	1.000	1.000	1.000	1.000	1.000	1.000	0.989	0.970	0.957	0.944
200	1.000	1.000	1.000	1.000	1.000	1.000	1.000	0.975	0.966	0.943
2000	1.000	1.000	1.000	1.000	1.000	0.999	0.987	0.974	0.947	0.931

Table 7.3 Average degree of linearity ($n = 14$, 1000000 problems)

	10%	20%	30%	40%	50%	60%	70%	80%	90%	100%
1	0.988	0.950	0.907	0.872	0.844	0.822	0.803	0.787	0.773	0.760
2	0.991	0.957	0.918	0.885	0.857	0.835	0.815	0.799	0.785	0.773
20	0.994	0.971	0.940	0.909	0.882	0.859	0.840	0.823	0.808	0.795
200	0.994	0.971	0.939	0.908	0.881	0.858	0.838	0.822	0.807	0.794
2000	0.994	0.971	0.938	0.908	0.881	0.858	0.838	0.821	0.807	0.794

- The minimum degrees of linearity can be found in very sparse problems with small entries.
- Problems with higher density have a lower degree of linearity.
- Problem examples giving values close to the worst case bound are extremely rare.

Since the degrees of linearity observed in the above experiments are far from the worst case 0.5, we tried to generate more instances in order to have a chance to get lower degrees. Table 7.4 shows some results suggesting again that worst case examples must be extremely rare. In addition, the table also displays the maximum integrality gap encountered for 3-dicycle relaxations. It should be difficult to generate small instances where the 3-dicycle bound exceeds the optimum by more than 5%.

The discussion of facet strengths in the preceding chapter showed that there are instances with $n = 8$ and trivial gap 1.538 (corresponding to the degree of linearity 0.650) and 3-dicycle gap 1.077. In our experiments we did not even come close to these values which are themselves still far away from the worst case. Probably such bad cases really only occur for large n and sparse digraphs (as supported by the theoretical results). Worst cases should be extremely rare and therefore it is no surprise that they did not come up in our experiments because we could generate only a very small fraction of possible problem instances.

Table 7.4 also gives the percentage of problems (BB) which required a branch-and-bound, i.e., for which the 3-dicycle relaxation did not give an integral solution. For larger problems with comparatively low degree of linearity a solution with the 3-dicycle LP only cannot be expected. The percentage of 81.4% for $n = 35$ is obviously too low, but caused by the fact that only very few instances could be solved.

Table 7.4 Computational results for random problems

n	Density	#problems	BB	min DoL	max DoL	3-cyc gap
10	1.0	10008189	0.1%	0.6889	1.0000	1.0323
20	0.1	8500349	0.6%	0.7647	1.0000	1.0357
20	0.5	327350	30.1%	0.7238	0.9333	1.0221
20	1.0	545636	63.1%	0.6737	0.8263	1.0365
30	0.1	1312510	26.4%	0.8269	1.0000	1.0217
30	0.2	88478	55.1%	0.8000	1.0000	1.0159
30	1.0	1974	99.8%	0.6713	0.7379	1.0471
35	1.0	118	81.4%	0.6756	0.7109	1.0446

7.4 Semidefinite Relaxations

Let $T_1,\ldots,T_m$, $m = n!$, denote all spanning acyclic tournaments of $D_n = (V_n, A_n)$ with their characteristic vectors χ^i. Throughout this monograph algorithms for solving the LOP to optimality were based on linear relaxations of

$$P^n_{LO} = \text{conv}(\{\chi^i \mid i = 1,\ldots,m\})$$

for computing upper bounds on the optimum objective function value, i.e., the bounds were obtained by solving linear programs of the form $\max\{c^T x \mid Ax \le b\}$ where $P^n_{LO} \subseteq \{x \mid Ax \le b\}$. We have experienced that the computation of these bounds is very time consuming and for difficult problems the upper bounds are fairly weak.

A more powerful alternative could be the employment of so-called *semidefinite relaxations* which require the solution of semidefinite programming problems. With symmetric (n,n)-matrices $C, A_1,\ldots,A_m$ and scalars $b_1,\ldots,b_m$ a semidefinite program (SDP) can be defined as

$$\min\{\langle C,X\rangle \mid \langle A_i,X\rangle = b_i, i = 1,\ldots,m, X \succeq 0\}.$$

Here $\langle C,X\rangle$ denotes the product $\sum_{i=1}^n \sum_{j=1}^n c_{ij}x_{ij}$ and $X \succeq 0$ denotes the requirement that X is a positive semidefinite matrix. Semidefinite programs can e.g. be solved by interior point methods. We cannot go into details on semidefinite programming here, but refer to the state-of-the-art survey [145].

There a several ways of coming up with a semidefinite relaxation for the LOP. A first one is the following where the central idea is that instead of P^n_{LO} we now consider

$$\mathcal{M}^n_{LO} = \text{conv}(\{\chi^i(\chi^i)^T \mid i = 1,\ldots,m\}).$$

This set lies in the space of symmetric (n,n)-matrices and in addition every matrix in $\mathcal{M}^n_{LO}$ is positive semidefinite. The additional property that we are interested only in matrices of rank 1 cannot be handled by SDP solvers and hence it is relaxed. Of course, every constraint for the original problem can be transformed into a constraint in the new space. The big advantage is that also quadratic constraints now become linear and can thus be employed in models. A disadvantage of this approach is that the dimension of the problem space is squared, i.e., in the case of the LOP increases from $O(n^2)$ to $O(n^4)$.

When using SDP relaxations it is useful to go from the 0/1-model to a ± 1-model. To this end we first replace the usual characteristic vectors χ^i for tournaments by ± 1-vectors ξ^i by setting

$$\xi^i = 2\chi^i - 1, i = 1, \dots, m.$$

Inequalities for the original characteristic vectors can easily be transformed into the new vectors. E.g., the 3-dicycle inequalities now read

$$-1 \le x_{ij} + x_{jk} - x_{ik} \le 1,$$

for all $i < j < k$. So essentially the model is not changed, but a major benefit is that now, in the semidefinite program, we can request in addition that the diagonal elements have value 1 as they correspond to squares of 1 or -1.

Instead of starting with a linear formulation of the LOP we could as well start with a quadratic formulation and then construct a semidefinite relaxation by going from vectors to matrices as described above. Such an approach was taken by Newman in [130]. It is based on a quadratic programming formulation of the LOP with ± 1-variables y_{ik} defined as

$$y_{ik} = \begin{cases} -1, & \text{the position of } i \text{ in the ordering is less than } k, \\ +1, & \text{the position of } i \text{ in the ordering is at least } k. \end{cases}$$

The formulation of Newman is

$$\max \sum_{(i,j) \in A_n} \sum_{1 \le h \le l \le n} \frac{1}{4} c_{ij} (y_{ih} - y_{i,h-1})(y_{jl} - y_{j,l-1})$$

$$(y_{ih} - y_{i,h-1})(y_{jl} - y_{j,l-1}) \ge 0, \quad i, j \in V_n, h, l = 1, \dots n,$$

$$y_{ih} y_{ih} = 1, \quad i \in V_n, h = 1, \dots n,$$

$$y_{i0} = -1, i \in V_n,$$

$$y_{in} = 1, \quad i \in V_n,$$

$$\sum_{i,j \in V_n} y_{i,\frac{n}{2}} y_{j,\frac{n}{2}} = 0,$$

$$y_{ih} \in \{-1, +1\}, i, h = 1, \dots, n.$$

The quality of the semidefinite relaxation based on this model is analyzed in [130]. Surprisingly, for the worst case examples of the 3-dicycle relaxation which give an integrality gap arbitrarily close to 2, the gap of this relaxation is only at most 1.64.

7.5 Context Independent Solvers

General purpose heuristics are based on models that treat the objective function evaluation as a black box, making the search algorithm context independent. The evaluation of the objective function can be seen as the call of an oracle which returns the objective value of a feasible solution. The solution algorithm itself does not know anything about the structure of this objective function and therefore cannot exploit structural properties. In particular, it is unknown to the solver whether the objective function is linear or nonlinear and therefore it possibly cannot apply the most effective search strategies for a given situation.

Meta-heuristics can be used to create solution procedures that are context independent. The original genetic algorithmic designs were based on this model. The advantage of this design is that the same solver can be applied to a wide variety of problems. The obvious disadvantage is that the solutions found by context independent solvers might be inferior to those of specialized procedures when both are allotted the same amount of computer effort (e.g. total search time).

Context independent solvers (also called general purpose or black box optimizers) based on meta-heuristics have found their home in commercial implementations. A standard evolutionary solver that is a context independent GA implementation is included in the Premium Solver Platform of Frontline Systems, Inc (www.frontsys.com). Opttek Systems, Inc (www.opttek.com) commercializes OptQuest, a context independent solver based on scatter search. Other GA-based commercial implementations of general purpose optimizers are Evolver by Palisade Corporation (www.palisade.com) and Pointer by Synaps, Inc. (www.synaps-inc.de).

In [26] a hybrid meta-heuristic for a class of problems based on a context independent paradigm is proposed. The method is restricted to those problems whose solutions are represented by permutations. This class includes a wide range of problems such as the traveling salesman problem, the quadratic assignment problem, various single machine sequencing problems, and the linear ordering problem, to mention only a few. The procedure is a combination of scatter search and of tabu search. The scatter search framework provides a means for diversifying the search throughout the exploration of the permutation solution space. Two improvement methods are used to intensify the search in promising regions of the solution space. Improved solutions are then used for combination purposes within the scatter search design. We briefly describe this design in the next paragraphs.

As described in Chapter 3, there are three elements that we need to define in any evolutionary method: a way to generate solutions, a way to combine solutions and a way to maintain a set (population) of solutions. The procedure in [26] follows the standard scatter search design to maintain the set of solutions (reference set). A generator of solutions (permutations), which focuses on diversification and not on the quality of the resulting solutions, is used at the beginning of the search to build the initial set P of solutions. The generator, proposed by Glover [72], uses a systematic approach to creating a diverse set of permutations. This contrasts with the typical GA approach of randomly generating an initial set of solutions from which to start the evolutionary search. In order to obtain a set of solutions of reasonable quality and diversity, an improvement method is applied to the solutions in P. The improvement method consists of two phases, a simple local search based on exchange moves and a tabu search. The TS is based on a short-term memory function and is applied only to the most promising solutions.

In order to design a context independent combination methodology that performs well across a wide collection of different problems, a set of ten combination methods is proposed (cm_1 to cm_{10}), from which one is probabilistically selected according to its performance in previous iterations. Solutions in the reference set are typically ordered according to their objective function value. So, the best solution is the first one in this set and the worst is the last one. A score is updated for every method as follows. If a solution obtained with combination method cm_i qualifies to be the jth member of the current reference set, then $b - j + 1$ is added to the score of cm_i. Therefore, combination methods that generate good solutions accumulate higher scores and are used more often.

In the experiments of [26] the authors considered four well known problems in which solutions are represented by permutations. We report in Table 7.5 the comparison between the proposed solver, called SS-TS, and two commercial packages, Frontline and Opttek, when solving the linear ordering and the traveling salesman problem. Table 7.5 reports the average percentaged deviation between the best solution obtained with each method and the optimal solution of 49 input-output instances of LOLIB (IO set) and 31 instances of TSPLIB (a library for the traveling salesman problem [161]). The table also shows the average CPU time in seconds.

Table 7.5 Comparison with commercial solvers

	OptQuest	Frontline	SS-TS
Linear Ordering Problem			
Deviation	8.5%	16.1%	0.0%
CPU time	301	300	25
Traveling Salesman Problem			
Deviation	311.4%	8.4%	5.7%
CPU time	5772	5628	23

Table 7.5 shows that the proposed scatter search with a tabu search improvement method yields higher quality solutions on average when compared to two commercially available software packages. To make a fair comparison, a fixed number of objective function evaluations has been set as a termination criterion for all procedures. We include the execution time to show the advantage of using a specialized code that does not include additional costly routines, such as those associated with graphical output or databases to store all visited solutions. It must be noted that, although we mentioned in previous chapters that the input-output instances of LOLIB are relatively easy to solve with the meta-heuristic methodologies, the solvers in Table 7.5 are context independent and they do not employ the knowledge, properties and structure of the problem as meta-heuristics can do.

As a final note in this section, we want to mention an interesting analysis that may permit to transfer solving methods from one problem to another one. In particular, Hernando et al. [83] introduce the concept of intersection between the rankings of solutions produced by combinatorial problems, and they study those based on permutations. Their application with the LOP and the TSP opens an interesting area for future research of black-box solvers.

The authors point out that if the solutions of an instance of a combinatorial optimization problem are sorted according to their objective function values, we can see the instances as (partial) rankings of the solutions of the search space. Working with specific problems, particularly, the linear ordering problem and the asymmetric traveling salesman problem, they show that they cannot generate the whole set of (partial) rankings of the solutions of the search space, but just a subset. Therefore they characterize the set of (partial) rankings each problem can generate, and then study the intersections between these problems: those rankings which can be generated by both the linear ordering problem and the symmetric/asymmetric traveling salesman problem, respectively. The fact of finding large intersections between problems can be useful to transfer heuristics from one problem to another one, or to define general heuristics that can be useful for more than one problem, in a step to create if not black-box solvers, methods that solve a class of problems.

7.6 Difficulty of LOP Instances

A problem instance is not in itself difficult, but it is difficult with respect to the solution algorithm. Clearly, for branch-and-bound or branch-and-cut, a problem should be difficult if the relaxations used do not provide good bounds.

Figures 7.2 and 7.3 display the correlation between CPU times for solving the problems of the previous section to optimality and the respective integrality gaps of the tournament and the 3-dicycle LP. Obviously, problems with higher gaps are more difficult for branch-and-cut algorithms.

From the perspective of heuristic and meta-heuristic methodologies we can differentiate three categories of methods:

a) constructive methods,
b) local search based methods,
c) population based methods.

An instance is easy to solve with a constructive method if it contains useful information associated with partial neighborhoods. In other words, if we have useful evaluations both for selecting a good candidate and for placing it in a good position to extend the partial solution under construction. In this case we say that the constructive process is guided by context information.

An instance is easy to solve with a local search based method if it contains useful information associated with complete neighborhoods. If the evaluation associated with the moves in a neighborhood permits us to discriminate among them, the local

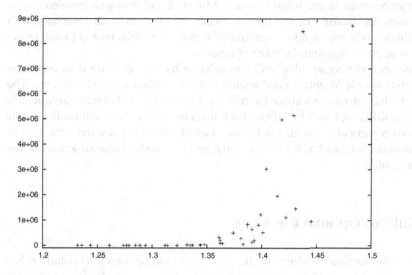

Fig. 7.2 Correlation between tournament gap and CPU time

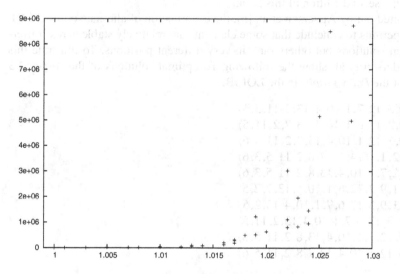

Fig. 7.3 Correlation between 3-dicycle gap and CPU time

search can progress in the solution space. Min-max and max-min problems typically present flat landscapes and do not contain information at all associated with the neighborhoods. Move values are usually 0 in these problems, and therefore decisions to select moves must be taken at random.

An instance is easy to solve with a population based procedure if its combination method is able to obtain good solutions when applied to good solutions. The conjecture that information about the relative desirability of alternative solutions is captured in different forms by different solutions motivated these approaches. Their success on a particular instance is based on its ability to capture the structures in good solutions responsible for their quality, and to transfer these structures to the combined solutions.

7.7 Multiple Optimal Rankings

In spite of its practical significance, the analysis of multiple optimal solutions has received little in the LOP literature. As far as we know, the main contribution [5] approaching this angle of the problem has been published this year 2021, in which we finish the second edition of this book.

As pointed out by Anderson et al. [5], the examination of alternate optimal LOP solutions permits to conclude that some elements are relatively stable across different optimal solutions but others rank in very different positions. To illustrate this point, Anderson et al. show the following 10 optimal solutions of the N-pal13 instance of the *Paley graphs* in the LOLIB.

1. $\langle 8,9,3,12,7,1,10,4,13,2,11,5,6 \rangle$
2. $\langle 8,9,3,12,6,1,10,4,13,7,2,11,5 \rangle$
3. $\langle 8,9,3,12,1,10,4,13,7,2,11,5,6 \rangle$
4. $\langle 9,12,1,10,4,13,7,8,2,11,5,3,6 \rangle$
5. $\langle 9,12,7,1,10,4,13,8,2,11,5,3,6 \rangle$
6. $\langle 8,11,9,3,12,6,1,10,4,13,7,2,5 \rangle$
7. $\langle 8,11,9,3,12,6,7,1,10,4,13,2,5 \rangle$
8. $\langle 8,9,3,12,6,7,1,10,4,13,2,11,5 \rangle$
9. $\langle 9,3,12,7,1,10,4,13,8,2,11,5,6 \rangle$
10. $\langle 9,3,12,1,10,4,13,7,8,2,11,5,6 \rangle$

It is clear from the optimal solutions above, that elements 8 and 9 are consistently in the first positions. However, there are other elements, such as 3 or 6 that rank in very different positions, depending on the optimal solution considered. In particular, in Solution 1, element 3 ranks in position 3, but in Solution 4 it ranks in position 12. Similarly, element 6 ranks in position 5 in Solution 2, but it ranks in the last position in many other optimal solutions. Considering that most of the solving methods proposed for the LOP only identified one optimal (or simply good) solution, we have to be careful when considering the individual positions of the elements, especially in some applications. That is especially true for those applications in which we want to obtain a ranking of the objects or elements for which we have pairwise comparisons, such as, for example, ranking in sports tournaments or lists of best universities. In these cases, the LOP seems not to be an appropriate model by itself, and require additional elements to disclose the most adequate ranking.

The existence of multiple optimal solutions is related to the weights in the data and an empirical analysis on the LOLIB [5] reveals that many instances have more than one optimal solution. The authors propose an exact method to obtain all the optimal solutions for small instances, and some of them (top 25 rankings) in medium size ones. Additionally, they identify some important (extreme) optimal solutions, that may help to obtain a final ranking. In particular, they consider the centroid of the set of optimal solutions, and the nearest and farthest solutions from the centroid. This study has obviously important implications in the application of the LOP and opens very interesting areas of research to enrich this model.

7.8 Sparse Problems

When most of the arcs of a LOP have zero weight, then it is preferrable to treat it as an acyclic subdigraph problem. As pointed out in the introduction ASP and LOP are trivially equivalent. If the set of arcs with positive weights is sparse, then a maximum weight acyclic subdigraph arc set will not necessarily be a tournament, but can be extended to a tournament.

In the sparse case, we do not optimize the objective function over P_{LO}^n but over the polytope $P_{AC}(D)$ instead, where $D = (V, A)$ is the subdigraph of D_n obtained by eliminating all arcs with weight zero. $P_{AC}(D)$ is defined as

$$P_{AC}(D) = \mathrm{conv}\{\chi^B \in \{0,1\}^A \mid B \text{ is an acyclic arc set in } A\}.$$

Further details on the ASP, in particular on the facet structure of $P_{AC}(D)$ can be found in [91]. Note that P_{LO}^n is a face of $P_{AC}(D_n)$ obtained by requiring equality in all 2-dicycle inequalities $x_{ij} + x_{ji} \leq 1$. Therefore, the facial structures of the two polytopes are closely related.

The canonical IP formulation of the ASP is

$$\max \sum_{(i,j) \in A} c_{ij} x_{ij}$$
$$x(C) \leq |C| - 1, \text{ for all dicycles } C \text{ in } A,$$
$$x_{ij} \in \{0,1\}, \ (i,j) \in A.$$

Note that in the case of the ASP we have to exclude all dicycles of length k, $k \geq 2$, because the digraph is not complete. Furthermore the tournament equations do not apply here. (For LOP instances in normal form, there are no 2-dicycles.)

If the integrality constraints are replaced by "$0 \leq x_{ij} \leq 1$", then we obtain the *dicycle relaxation* of the ASP. The separation problem for dicycle inequalities can

easily be solved in polynomial time, as the following observation shows. Let x^* be
the current fractional LP solution and define $y^* = 1 - x^*$. Then $x^*(C) \leq |C| - 1$ if
and only if $y^*(C) \geq 1$. So, for separating dicycle inequalites, the shortest dicycle
in A with arc weights given by y^* is computed (using shortest path techniques). If
the shortest such diycle has length greater than or equal 1, then no dicycle inequality
is violated, otherwise this dicycle yields a cut.

Therefore, the dicycle relaxation can be solved in polynomial time and can serve
as a basis for branch-and-bound and branch-and-cut algorithms which are designed
analogously as for the LOP.

Note that if we add the inequality system

$$x_e - x_f + x_g \leq 1, \text{ for all arcs } e, f, g \in A \text{ and } V(f) \subseteq V(\{e, g\}),$$

then we obtain an IP formulation of the node induced acyclic subdigraph problem
which can serve as a basis for branch-and-cut algorithms.

7.9 A Simple Dual Heuristic

Let a LOP in normal form be given. Denote its optimum solution by c_{opt} and let Q
denote the sum of all arc weights.

Let B be some dicycle in A_n and define

$$B_{\text{min}} = \min\{c_{ij} \mid (i, j) \in B\}.$$

Then $B_{\text{min}} \geq 0$ and obviously $c_{\text{opt}} \leq Q - B_{\text{min}}$. We modify the objective function by
setting

$$c'_{ij} = \begin{cases} c_{ij} - B_{\text{min}}, & (i, j) \in B, \\ c_{ij}, & \text{otherwise.} \end{cases}$$

With another dicycle B' and $B'_{\text{min}} = \min\{c'_{ij} \mid (i, j) \in B'\}$ we can improve the upper
bound to $c_{\text{opt}} \leq Q - B_{\text{min}} - B'_{\text{min}}$.

Of course, only dicycles with strictly positive minimum arc weight are of interest.
We can iterate this procedure as long as we find such dicycles and get an upper
bound for the optimum solution of the LOP.

DualHeuristic(C)

(1) Set $UB = Q$ and $c'_{ij} = c_{ij}$, for all $(i,j) \in A_n$.
(2) While $A_n \setminus \{(i,j) \mid c'_{ij} = 0\}$ is not acyclic:

(2.1) Let B be a dicycle in $A_n \setminus \{(i,j) \mid c'_{ij} = 0\}$.
(2.2) Set $B_{\min} = \min\{c'_{ij} \mid (i,j) \in B\}$.
(2.3) Set $UB = UB - B_{\min}$.
(2.4) For every $(i,j) \in B$ set $c'_{ij} = c'_{ij} - B_{\min}$.

Surprisingly, this heuristic yields fairly reasonable upper bounds. Table 7.6 gives the average percentage of this bound w.r.t the best known lower and upper bounds. We also list the percentage of this bound in terms of the sum of all objective function coefficients.

Table 7.6 Comparison of simple upper bounds

Problem class	% Sum	%best UB	%best LB
IO	96.01	100.46	100.46
Random A1	70.38	101.62	114.77
Random A2	94.46	101.29	101.55
Random B	76.48	105.07	105.73
MB	91.11	100.81	100.81
XLOLIB	85.16	106.78	109.66
SGB	73.16	100.89	100.89
Spec	85.59	104.57	106.90

Thus, in the case that LP bounds cannot be computed, these bounds provide useful information for assessing heuristics.

Now consider the following linear program.

$$\max \sum_{(i,j) \in A_n} c_{ij} x_{ij}$$

$$x(C) \leq |C| - 1, \text{ for all dicycles } C \text{ in } A_n, |C| \geq 2,$$

$$x_{ij} \leq 1, \ (i,j) \in A_n,$$

$$x_{ij} \geq 0, \ (i,j) \in A_n.$$

For a LOP in normal form, its optimum value is equal to the value of the 3-dicycle relaxation. The dual of this program is

$$\min \sum_{B \text{ dicycle in } A_n} (|B| - 1)y_B + \sum_{(i,j) \in A_n} z_{ij}$$

$$\sum_{\substack{B \text{ dicycle in } A_n \\ (i,j) \in B}} y_B + z_{ij} \geq c_{ij}, \text{ for all } (i,j) \in A_n,$$

$$y_B \geq 0, \text{ for all dicycles } B \text{ in } A_n,$$

$$z_{ij} \geq 0, (i,j) \in A_n.$$

Let $\mathscr{B}$ denote the set of dicycles used in the dual heuristic and set the dicycle variables

$$y_B = \begin{cases} B_{\min}, & B \in \mathscr{B}, \\ 0, & \text{otherwise}, \end{cases}$$

and

$$z_{ij} = c_{ij} - \sum_{B \in \mathscr{B}, (i,j) \in B} y_B.$$

Then this setting of variables is feasible for the dual program and

$$Q - \sum_{B \in \mathscr{B}} B_{\min} = \sum_{(i,j) \in A_n} \left(z_{ij} + \sum_{B \in \mathscr{B}, (i,j) \in B} y_B \right) - \sum_{B \in \mathscr{B}} B_{\min}$$

$$= \sum_{(i,j) \in A_n} z_{ij} + \sum_{B \in \mathscr{B}} (|B| - 1)y_B.$$

So the heuristic can be interpreted as finding a feasible solution for the dual of the 3-dicycle relaxation. Therefore it gives an upper bound on the value of the 3-dicycle relaxation and thus an upper bound on the LOP.

7.10 Future Research on the LOP

The current state-of-the-art of solution algorithms for the LOP seems not really satisfactory. Whereas very good feasible solutions can be found for large problems using the comprehensive toolbox of heuristics, fairly small problems can still be difficult for exact algorithms and cannot be solved to optimality in reasonable time. So there is definite need for further research on algorithms, mainly for exact algorithms but also for heuristics. We expect progress on the following topics.

Black Box Solvers

As we have seen in the computational experiments shown in Section 7.5, the quality of the solutions obtained with context independent or black box solvers is still moderate. Although it is expected that general purpose methods obtain medium quality solutions, as compared with the high-quality obtained with the specialized methods, the truth is that these generic methods need further development. Considering that most commercial solvers are based on this technology, we should study their associated models and heuristics algorithms to achieve better results.

Solution of the 3-Dicycle Relaxation

The determination of this bound is at the core of exact algorithms. The solution time needed is still not satisfactory. Possibly a combination of primal and dual methods, i.e., cutting plane and bundle approaches bears the potential for improvement.

Separation Algorithms

Besides the trivial 3-dicycle enumeration and the separation of certain Möbius ladders as mod-2 inequalities, there are no other exact separation algorithms. It is known that the separation of k-fences is NP-hard. As the computation for small polytopes shows there is a wealth of further facet defining inequalities the structure of which is unexplored. So there is room for theoretical research on the complexity of separation and the development of further exact algorithms.

Local Cuts and Target Cuts

Local cuts and target cuts offer the possibility of generating cutting planes for large problems and generating cuts which are different from the ones used in separation procedures so far. It could be promising to study their effect.

Branch-and-Bound Revisited

Most of the present day branching schemes are binary, i.e., some variable is fixed to 0 or to 1 and the corresponding two subproblems are created. The effect of just changing one variable is small and it could be advantageous to employ more complicated schemes.

An obvious possibility is to branch by fixing nodes at certain positions of the ordering. Fixing a node at a position has significantly more consequences than just fixing one variable. Furthermore, branching on inequalities could be an option and parallelization is always applicable to branch-and-bound.

Different Models

All computational methods for solving the LOP are based on the binary variables x_{ij} where

$$x_{ij} = \begin{cases} 1, & \text{if } i \text{ is before } j \text{ in the ordering,} \\ 0, & \text{otherwise.} \end{cases}$$

But other modeling approaches could be explored as well. E.g., one could use binary variables x_{ijk} for three nodes i, j, k defined as

$$x_{ijk} = \begin{cases} 1, & \text{if } j \text{ is after } i \text{ and before } k \text{ in the ordering,} \\ 0, & \text{otherwise.} \end{cases}$$

An IP formulation of the LOP with these variables is

$$\max \sum w_{ijk} x_{ijk}$$
$$x_{ijk} + x_{ikj} + x_{jik} + x_{jki} + x_{kij} + x_{kji} = 1$$
$$x_{ijk} + x_{ikj} + x_{kij} - x_{ijl} - x_{ilj} - x_{lij} = 0$$
$$x_{ijk} \in \{0,1\}$$

where

$$w_{ijk} = \frac{c_{ij} + c_{ik} + c_{kj}}{n-2},$$

if c denotes the original objective function. The canonical LP relaxation gives the 3-dicycle bound as the standard formulation with 2-index variables.

For $n = 4$ the convex hull of feasible 0/1 vectors has further facets defined by the inequalities

$$x_{ijk} - x_{ijl} - x_{ljk} \leq 0$$
$$x_{ijk} - x_{ijl} - x_{lik} - x_{ilk} \leq 0$$
$$x_{ijk} - x_{ikl} - x_{ljk} - x_{ilk} \leq 0$$
$$x_{ijk} + x_{lkj} - x_{ikl} - x_{ilk} - x_{lji} - x_{lij} \leq 0$$

for pairwise distinct i, j, k, l.

If all these inequalities are added, then the LP relaxation for $n = 6$ is integral, i.e., all 3-fence and Möbius ladder inequalities are implied.

The exploration of alternative models could possibly lead to more effective algorithms.

General Integer Programming

Since we have difficult problem instances already for small values of n, the study of general integer programming approaches like lift-and-project could be worthwhile.

Determination of Optimum Solutions Using SDP

In theory, semidefinite programming bounds are stronger than linear programming bounds. First experiments with quadratic models and semidefinite relaxations support this. On the other hand, the problem dimension is drastically increased. But, in the range of a few hundred nodes, this could be manageable and there should be a thorough investigation of the power of semidefinite relaxations in the context of the LOP. First experiments on small problems reveal that the 3-dicycle bound seems to be easily improvable by employing SDP.

7.11 Extensions of the MDP

In the seminal papers by Glover et al. [67] in 1995, simple heuristics were proposed for the MDP. The authors pointed out that different versions of this problem may include additional constraints, and their objective was to design heuristics with simple moves for transitioning from one solution to another to allow them to be adapted to multiple settings. In spite of this early interest in constrained models, the extensive literature on diversity problems is mostly based on selecting a fixed number of points, m, avoiding the use of additional constraints. In this section we review the few developments published on constrained versions of diversity problems.

Rosenkrantz et al. [149] introduced several diversity models constrained in terms of cost and capacity, motivated by their practical applications in facility location. For example, the location of undesirable or hazardous facilities, such as waste sites or nuclear plants, requires their dispersion while satisfying a certain total demand. Facilities have a capacity to provide a service in systems that require an overall demand, and it is clear in practical terms that they have an associated setup or operational cost, which makes appropriate to consider a certain limit in the total expenses generated. As stated by the authors, "these practical aspects add a new dimension to the conventional dispersion problem". Classical models, such as the MaxSum (MDP) or MaxMin (MMDP), indirectly address the problem requirements by considering a prefixed number of facilities (i.e., the number of points to be selected is an input to the problem). However, this simplification is not realistic in many settings. The first model proposed by these authors is the *Capacitated Dispersion Problem* (CDP) in which a capacity constraint is included in the Max-Min diversity problem.

Given a set of n potential facilities V connected by edges (links) in E, the CDP consists in finding a subset P of V satisfying the capacity constraint, so that the minimum distance among the sites in P is maximized. Let B be the minimum required capacity (level of service). Then, for each site $i \in V$, we define $c_i \geq 0$ as its capacity. Let $d_{ij} \geq 0$ be the distance between sites i and j. The mathematical programming model for CDP is based on the standard binary variables x_i that take the value 1 if site i is selected and 0 otherwise. Then, it can be stated as follows:

$$\max \min_{\substack{i,j \in V, i \neq j \\ x_i = x_j = 1}} d_{ij} x_i x_j$$

$$\sum_{i=1}^{n} c_i x_i \geq B$$

$$x_i \in \{0,1\}, \ i = 1, \dots, n.$$

The main drawback of this model is the non-linearity of the objective function, which prevents the use of standard mixed integer programming solvers. However, this quadratic problem can be reformulated as the following integer linear programming model [121]:

$$\max t$$

$$\sum_{i=1}^{n} c_i x_i \geq B$$

$$t \leq d_{ij} + U(2 - x_i - x_j) \qquad \forall\, 1 \leq i < j \leq n$$

$$x_i \in \{0,1\}, \ i = 1, \dots, n.$$

The upper bound U on the distances values guarantees that t is the minimum distance among the selected sites.

Rosenkrantz et al. [149] proposed a heuristic with performance guarantee to solve this NP-hard problem. The authors proved that on instances with inter-objects distances satisfying the triangle inequality, their heuristic has a performance guarantee of 2. Although no empirical results or experiments are reported, the theoretical study also concludes that their heuristic running time is $O(n^2 log n)$. Martí et al. [121] recently approached the CDP from a practical perspective. In particular, the authors propose heuristic algorithms based on state-of-the-art metaheuristic methodologies such as GRASP and VND with no performance guarantee, but their experimental results show that they statistically perform very well when solving a large set of instances.

Martínez-Gavara et al. [122] target a more realistic variant, called the *Generalized Dispersion Problem*, in which there are two constraints, capacity and cost, which are common elements in many real problems. For example, in the location of similar facilities, the selection of a subset of sites usually deals more with the total storage capacity and expenses rather than with a fixed number of sites.

Consider for example the location problem of a company that operates under the franchise business model, and wants to expand their business opening new stores in a given city, where they do not have previously any presence. If we applied the MDP model to solve this problem, we would only consider the inter-distance maximization between the selected points, and we would be ignoring the cost a_i or capacity c_i associated to each site i. This is why we say that the MDP is unrealistic in some situations like this one. The application of the generalized dispersion model on the other hand, permits to define a minimum number of customers B to provide service, and a maximum number of expenses, K. Therefore, instead of considering to select a pre-established number of locations, as the MDP model does, the generalized model selects as many sites as needed to achieve the capacity required without exceeding the available budget. Additionally, this model also considers the maximization of inter-distances between selected sites to avoid local competition, making it specially adequate for this type of problem. It is thus unquestionable that in terms of solving a real location problem, the generalized model reflects in a better way the particularities of the real problem than the classic MDP.

The GDP can be modeled in the same way than the CDP as follows:

$$\max \min_{\substack{i,j \in V, i \neq j \\ x_i = x_j = 1}} d_{ij} x_i x_j$$

$$\sum_{i=1}^{n} c_i x_i \geq B$$

$$\sum_{i=1}^{n} a_i x_i \leq K$$

$$x_i \in \{0, 1\}, \ i = 1, \ldots, n.$$

Martínez-Gavara et al. [122] also considered a more realistic variant of the GDP, already proposed in [149], in which the total cost is a linear function of a fixed setup cost and a variable cost reflecting the amount of material stored or number of users served, depending on the application. This problem can be formulated in mathematical terms using the same notation than above, and adding the integer variable t_i as the units of material stored, and their cost (per unit) b_i in site i, for $i \in V$. Note that the stored material t_i cannot exceed the capacity of the site, c_i, and t_i should be 0 if the site i is not selected. We model it in the new formulation as the constraint $t_i \leq c_i x_i$, for each i in V. We called this model the Generalized Dispersion Problem with variable costs, and it is formulated as follows:

$$\max \quad \min_{i,j \in M \subset V} \quad d_{ij} x_i x_j$$

$$\text{s.t.} \quad \sum_{i=1}^{n} t_i \geq B$$

$$\sum_{i=1}^{n} a_i x_i + b_i t_i \leq K$$

$$t_i \leq c_i x_i \qquad \forall i \in V$$

$$t_i \in \mathbb{Z}, x_i \in \{0,1\} \qquad \forall i \in V$$

The authors proposed both efficient heuristics to target large instances, and linearizations of this model to obtain optimal solutions with CPLEX on the small and medium size instances. These constrained problems clearly open new possibilities to apply diversity maximization to model real problems, specially in location theory. We can anticipate an important development in this family of problems in the next years.

References

1. ACHATZ, H., KLEINSCHMIDT P. AND LAMBSDORFF, J.: *Der Corruption Perceptions Index und das Linear Ordering Problem*, ORNews **26** (2006), 10–12.
2. ACHTERBERG, T.: *SCIP: Solving constraint integer programs*, Math. Programming Computation **1** (2009), 1–41.
3. ADOLPHSON, D. AND HU, T.C.: *Optimal linear ordering*, SIAM J. on Applied Mathematics **25** (1973), 403–423.
4. AARTS, E. AND LENSTRA, J.K.: *Local search in combinatorial optimization*, Princenton University Press, 2003.
5. ANDERSON, P.E., CHARTIER, T.P., LANGVILLE, A.N. AND PEDINGS-BEHLING, K.E.: *Fairness and the set of optimal rankings for the linear ordering problem*, Optimization and Engineering. https://doi.org/10.1007/s11081-021-09650-y, 2021.
6. APPLEGATE, D.L., BIXBY, E.B., CHVÁTAL, V. AND COOK, W.J.: *The traveling salesman problem: A computational study*, Princeton University Press, 2006.
7. AUJAC, H.: *La hiérarchie des industries dans un tableau des echanges industriels*, Rev. Economique **2** (1960).
8. ARINGHIERI, R. AND CORDONE, R.: *Comparing local search metaheuristics for the maximum diversity problem*, Journal of the Operational Research Society **62** (2011), 266-280.
9. ARINGHIERI, R., CORDONE, R. AND GROSSO, A.: *Construction and improvement algorithms for dispersion problems*, European Journal of Operation Research **242** (2015), 21-33.
10. BALUJA, S.: *An empirical comparison of 7 iterative evolutionary function optimization heuristics*, School of computer science, Carnegie Mellon University, 1995.
11. BEASLEY, J. E.: *OR-Library: Distributing Test Problems by Electronic Mail*, Journal of the Operational Research Society **41** (1990) 1069–1072.
12. BECKER, O.: *Das Helmstädtersche Reihenfolgeproblem – die Effizienz verschiedener Näherungsverfahren*, in: *Computer uses in the social sciences*, Bericht einer Working Conference des Inst. f. höh. Studien u. wiss. Forsch., Wien, 1967.
13. BÉKÉSI, J., GALAMBOS, G., OSWALD, M. AND REINELT, G.: *Comparison of approaches for solving coupled task problems*, Technical Report, U Heidelberg, 2008.
14. BERTACCO, L., BRUNETTA, L., AND FISCHETTI, M.: *The linear ordering problem with cumulative costs*, European Journal of Operational Research **189** (2008), 1345–1357.
15. BHANDARKAR, S.M. AND CHIRRAVURI, S.: *A study of massively parallel simulated annealing algorithms for chromosome reconstruction via clone ordering*, Parallel Algorithms and Applications **9** (1996), 67–89.
16. BOENCHENDORF, K.: *Reihenfolgenprobleme / Mean-flow-time sequencing*, Mathematical Systems in Economics 74, Verlagsgruppe Athenäum, Hain, Scriptor, 1982.
17. BOESE, K.D., KAHNG, A.B. AND MUDDU S.: *A new adaptive multi-start technique for combinatorial global optimisation*, Operations Research Letters **16** (1994), 103–113.

18. BOLOTASHVILI, G., KOVALEV, M. AND GIRLICH, E.: *New facets of the linear ordering polytope*, Siam J. Discrete Math. **12** (1999), 326–336.

19. BOWMAN, V.J.: *Permutation polyhedra*, SIAM J. Appl. Math. **22** (1972), 580–589.

20. BRIMBERG, J., MLADENOVIĆ, N., UROŠEVIĆ, D. AND NGAI, E.: *Variable neighborhood search for the heaviest -subgraph*, Computers and Operations Research **36** (2009) 2885–2891.

21. BUCHHEIM, CH., LIERS, F. AND OSWALD, M.: *Local cuts revisited*, Operations Research Letters **36** (2008), 430–433.

22. BUCHHEIM, CH., WIEGELE, A. AND ZHENG, K.: *Exact algorithms for the quadratic linear ordering problem*, INFORMS J. on Computing **22** (2010), 168–177.

23. BURKARD, R.E. AND FINCKE, U.: *Probabilistic asymptotic properties of some combinatorial optimization problems*, Discrete Appl. Math. **12** (1985), 21–29.

24. CAMPOS, V., GLOVER, F. LAGUNA, M. AND MARTÍ, R.: *An experimental evaluation of a scatter search for the linear ordering problem*, J. of Global Optimization **21** (2001), 397–414.

25. CAMPOS, V., LAGUNA, M. AND MARTÍ, R.: *Scatter search for the linear ordering problem*, in: Corne, D., Dorigo, M. and Glover, F. (eds.), *New Ideas in Optimisation*, McGraw-Hill, 1999, 331–341.

26. CAMPOS, V., LAGUNA, M. AND MARTÍ, R.: *Context-independent scatter and tabu search for permutation problems*, INFORMS J. on Computing **17** (2005), 111–122.

27. CAPRARA, A. AND FISCHETTI, M.: $\{0, \frac{1}{2}\}$-*Chvátal-Gomory cuts*, Math. Programming **74** (1996), 221–235.

28. CAPRARA, A., FISCHETTI, M. AND LETCHFORD, A.N.: *On the separation of maximally violated mod-k cuts*, Math. Programming **87** (2000), 37–56.

29. CEBERIO, J., MENDIBURU, A. AND LOZANO, J.A.: *The linear ordering problem revisited*, European J. of Operational Research **241** (2015), 686–696.

30. CEBERIO, J., MENDIBURU, A. AND LOZANO, J.A.: *Are We Generating Instances Uniformly at Random?*, 2017 IEEE Congress on Evolutionary Computation (CEC-2017), 1645–1651, Donostia/San Sebastian, Spain, 5-8 June 2017.

31. CHANAS, S. AND KOBYLANSKI, P.: *A new heuristic algorithm solving the linear ordering problem*, Computational Optimization and Applications **6** (1996), 191–205.

32. CHARON, I. AND HUDRY, O.: *Lamarckian genetic algorithms applied to the aggregation of preferences*, Annals of Operations Research **80** (1998), 281–297.

33. CHARON, I. AND HUDRY, O.: *The noising methods: A generalization of some metaheuristics*, European J. of Operational Research **135** (2001), 86–101.

34. CHARON, I. AND HUDRY, O.: *A branch-and-bound algorithm to solve the linear ordering problem for weighted tournaments*, Discrete Applied Mathematics **156** (2006), 2097–2116. `http://www.enst.fr/~charon/tournament/median.html`

35. CHARON, I. AND HUDRY, O.: *A survey on the linear ordering problem for weighted or unweighted tournaments*, 4OR **5** (2007), 5–60.

36. CHARON, I. AND HUDRY, O.: *An updated survey on the linear ordering problem for weighted or unweighted tournaments*, Annals of Operations Research **175** (2009), 107–158.

37. CHENERY, H.B. AND WATANABE, T.: *International comparisons of the structure of production*, Econometrica **26** (1958), 487–521.

38. CHIARINI, B., CHAOVALITWONGSE, W. AND PARDALOS, P.M.: *A new algorithm for the triangulation of input-output tables*, in: Pardalos, P.M., Migdala, A. and Baourakis, G. (eds.), *Supply chain and finance*, World Scientific, 2004, 254–273.

39. CHRISTOF, T.: *Low-dimensional 0/1-polytopes and branch-and-cut in combinatorial optimization*, Shaker, 1997.

40. CHRISTOF, T. AND REINELT, G.: *Combinatorial optimization and small polytopes*, Top **4** (1996), 1–64.

41. COELLO, C. A., VAN VELDHUIZEN, D. A. AND LAMONT, G. B.: *Evolutionary algorithms for solving multi-objective problems*, Kluwer Academic / Plenum Publishers, 2002.

42. CONDORCET, M.J.A.N.: *Essai sur l'application de l'analyse à la probabilité des décisions rendues à la pluralité des voix*, Paris, 1785.

43. DAKIN, R.J.: *A tree search algorithm for mixed integer programming problems*, Computer Journal **8** (1965), 250–255.
44. DAVIS, L.: *Handbook of genetic algorithms*, International Thomson Computer Press, 1996.
45. DE CANI, J.S.: *Maximum likelihood paired comparison ranking by linear programming*, Biometrika **56** (1969), 537–545.
46. DE CANI, J.S.: *A branch & bound algorithm for maximum likelihood paired comparison ranking*, Biometrika **59** (1972), 131–135.
47. DOIGNON, J.-P., FIORINI, S. AND JORET, G.: *Facets of the linear ordering polytope: A unification for the fence family through weighted graphs*, J. of Math. Psychology **50** (2006), 251–262.
48. DREO, J., PETROWSKI, A., SIARRY, P. AND TAILLARD, E.: *Metaheuristics for hard optimization*, Springer, 2006.
49. DRIDI, T.: *Sur les distribution binaires associées à des distributions ordinales*, Math. Sci. hum. **69** (1980), 15–31.
50. DUARTE, A., LAGUNA, M. AND MARTÍ, R.: *Tabu search for the linear ordering problem with cumulative costs*, Computational Optimization and Applications **48**(3) (2009), 697-715.
51. DUARTE, A. AND MARTÍ, R.: *Tabu Search and GRASP for the Maximum Diversity Problem*, European Journal of Operational Research **178** (2007), 71-84.
52. DUARTE, A., LAGUNA, M. AND MARTÍ, R.: *MetaHeuristics for Business Analytics. A Decision Modeling Approach*, EURO Advanced Tutorials on Operational Research, Springer, (2018).
53. DUARTE, A., J. SÁNCHEZ-ORO, M. RESENDE, F. GLOVER AND MARTÍ, R.: *GRASP with Exterior Path Relinking for Differential Dispersion Minimization*, Information Sciences **296**, (2015), 46-60.
54. KUBY, M.J.: *Programming models for facility dispersion: the p-dispersion and maxisum dispersion problems*, Mathematical and Computer Modelling **792**, (1988), 10.
55. ERKUT, E. AND NEUMAN, S.: *Analytical models for locating undesirable facilities*, European Journal of Operational Research **40** (1989), 275–291.
56. FEO, T. AND RESENDE, M.G.C.: *A probabilistic heuristic for a computationally difficult set covering problem*, Operations Research Letters **8** (1989), 67–71.
57. FEO, T. AND RESENDE, M.G.C.: *Greedy randomized adaptive search procedures*, J. of Global Optimization **2** (1995), 1–27.
58. FIORINI, S.: *{0,1/2}-cuts and the linear ordering problem: Surfaces that define facets*, SIAM J. on Discrete Mathematics **20** (2006), 893–912.
59. FISHBURN, P.C.: *Binary probabilities induced by rankings*, SIAM J. Disc. Math. **3** (1990), 478–488.
60. FISHBURN, P.C.: *Induced binary probabilities and the linear ordering polytope: a status report*, Math. Soc. Sciences **23** (1992), 67–80.
61. GALLEGO, M., DUARTE, A., LAGUNA, M., AND MARTÍ, R.: *Hybrid heuristics for the maximum diversity problem*, Computational Optimization and Applications **44**(3) (2009), 411-426.
62. GARCIA, C.G., PÉREZ-BRITO, D., CAMPOS, V. AND MARTÍ, R.: *Variable neighborhood search for the linear ordering problem*, Computers and Operations Research **33** (2006), 3549–3565.
63. GARCIA, E., CEBERIO, J. AND LOZANO, J.A.: *Hybrid Heuristics for the Linear Ordering Problem*, 2019 IEEE Congress on Evolutionary Computation (CEC-2019), 1431-1438, Wellington, New Zeland.
64. GAREY, M.R. AND JOHNSON, D.S.: *Computers and intractability: A guide to the theory of NP-completeness*, Freeman, 1979.
65. GILBOA, I.: *A necessary but insufficient condition for the stochastic binary choice problem*, J. of Math. Psychology **34** (1990), 371–392.
66. GHOSH, J.B.: *Computational aspects of the maximum diversity problem*, Operations Research Letters **19** (1996), 175-181.
67. GLOVER, F., KUO, C.C. AND DHIR, K.S.: *A discrete optimization model for preserving biological diversity*, Applied Mathematical Modeling **19** (1995), 696-701.

68. GLOVER, F., KUO, C.C., AND DHIR, K.S.: *Heuristic algorithms for the maximum diversity problem*, Journal of Information and Optimization Sciences **19(1)** (1998), 109-132.

69. GLOVER, F. AND LAGUNA, M.: *Tabu search*, Kluwer Academic Publishers, 1997.

70. GLOVER, F.: *Heuristics for integer programming using surrogate constraints*, Decision Sciences **8** (1977), 371–392.

71. GLOVER, F.: *Future paths for integer programming and links to artificial intelligence*, Computers and Operations Research **13** (1986), 533–549.

72. GLOVER, F.: *A template for scatter search and path relinking*, in: Hao, J.-K., Lutton, E., Ronald, E., Schoenauer, M. and Snyers, D. (eds.), *Artificial Evolution*, LNCS 1363, Springer, 1998, 13–54.

73. GLOVER, F., KLASTORIN, T. AND KLINGMAN, D.: *Optimal weighted ancestry relationships*, Management Science **20** (1974), 1190–1193.

74. GOEMANS, M.X.: *Worst-case comparision of valid inequalities for the TSP*, Mathematical Programming **69** (1995), 335–349.

75. GOEMANS, M.X. AND HALL, L.A.: *The strongest facets of the acyclic subgraph polytope are unknown*, in: *Proc. of the 5th Int. IPCO Conference*, LNCS 1084, Springer, 1996, 415–429.

76. GRÖTSCHEL, M., JÜNGER, M. AND REINELT, G.: *A cutting plane algorithm for the linear ordering problem*, Operations Research **32** (1984), 1195–1220.

77. GRÖTSCHEL, M., JÜNGER, M. AND REINELT, G.: *Facets of the linear ordering polytope*, Math. Programming **33** (1985), 43–60.

78. GRÖTSCHEL, M., LOVASZ, L. AND SCHRIJVER, A.: *Geometric algorithms and combinatorial optimization*, Springer, 1988.

79. GRUNDEL, D.A. AND JEFFCOAT, D.E.: *Formulation and solution of the target visitation problem*, in: *Proc. of the AIAA First Intelligent System Technical Conference*, Chicago, 2004.

80. HANSEN, P. AND MLADENOVIC, N.: *Variable neighborhood search*, in: Glover, F. and Kochenberger, G. (eds.) *Handbook of Metaheuristics* (2003), 145–184.

81. HELLMICH, K.: *Ökonomische Triangulierung*, Rechenzentrum Graz, Heft 54, 1970.

82. HELMSTÄDTER, E.: *Die Dreiecksform der Input-Output-Matrix und ihre möglichen Wandlungen im Wachstumsprozeß*, in: Neumark, F. (ed.): *Strukturwandlungen einer wachsenden Wirtschaft* Schriften des Vereins für Socialpolitik, NF Band 30/II (1964), 1059–1063.

83. HERNANDO, L., MENDIBURU, A. AND LOZANO, J.A.: *Characterising the rankings produced by combinatorial optimisation problems and finding their intersections*, GECCO '19: Proceedings of the Genetic and Evolutionary Computation Conference, 266-273, Prague, Czech Republic, July 13-17, 2019.

84. HERNANDO, L., MENDIBURU, A. AND LOZANO, J.A.: *Journey to the Center of the Linear Ordering Problem*, GECCO '20: Proceedings of the 2020 Genetic and Evolutionary Computation Conference, 201-209, Cancun, Mexico, July 8-12, 2020.

85. HERTZ,A. AND WIDMER, M.: *Guidelines for the use of meta-heuristics in combinatorial optimization*, European J. of Operational Research **151** (2003), 247–252.

86. HICKERNELL, F.J. AND YUAN, Y.: *A simple multistart algorithm for global optimization*, OR Transactions **1(2)** (1997).

87. HOLLAND, J.H.: *Adaptation in natural and artificial systems*, University of Michigan Press, 1975.

88. HOLUB, H.-W. AND SCHNABL, H.: *Input-Output-Rechnung: Input-Output-Tabellen*, Oldenbourg, 1982.

89. HUANG, G. AND LIM, A.: *Designing a hybrid genetic algorithm for the linear ordering problem*, in: Cantu-Paz, E. et al. (eds.): *Proc. of Genetic and Evolutionary Computation – GECCO 2003*, LNCS 2723, Springer, 2003, 1053–1064.

90. JOHNSON, D., ARAGON, C.R., MCGEOCH, L.A. AND SCHEVON, C.: *Optimization by simulated annealing: An experimental evaluation; Part I, graph partitioning*, Operations Research **37** (1989), 865–892.

91. JÜNGER, M.: *Polyhedral combinatorics and the acyclic subdigraph problem*, Heldermann, Berlin, 1985.

92. JÜNGER, M. AND MUTZEL, P.: *2-layer straightline crossing minimization: Performance of exact and heuristic algorithms*, J. of Graph Algorithms and Applications **1** (1997), 1–25.
93. JÜNGER, M. AND THIENEL, S.: *The ABACUS System for branch-and-cut-and-price algorithms in integer programming and combinatorial optimization*, Software: Practice and Experience **30** (2000), 1325–1352.
    http://www.informatik.uni-koeln.de/abacus
94. KAAS, R.: *A branch and bound algorithm for the acyclic subgraph problem*, European J. of Operational Research **8** (1981), 355–362.
95. KANG, S.: *Linear ordering and application to placement*, in: *Proc. of the 20th Annual ACM IEEE Design Automation Conference*, Miami Beach, Floria, USA (1983), 457–464.
96. KEMENY, J.G.: *Mathematics without numbers*, Daedalus **88** (1959), 577–591.
97. KERNIGHAN, B.W. AND LIN, S.: *An efficient heuristic procedure for partitioning graphs*, Bell Systems Technical Journal **49** (1979) 291–308.
98. KIRKPATRICK, S., GELATT, C.D. AND VECCHI, M.P.: *Optimization by simulated annealing*, Science **222** (1983), 671–680.
99. KNUTH, D.E.: *The Stanford GraphBase: A platform for combinatorial computing*, Addison-Wesley, 1993.
100. KOPPEN, M.: *Random utility representation of binary choice probabilities: critical graphs yielding critical necessary conditions*, J. of Math. Psychology **39** (1995), 21–39.
101. KORTE, B. AND OBERHOFER, W.: *Zwei Algorithmen zur Lösung eines komplexen Reihenfolgeproblems*, Unternehmensforschung **12** (1968), 217–231.
102. KORTE, B. AND OBERHOFER, W.: *Zur Triangulation von Input-Output-Matrizen*, Jahrbücher für Nationalökonomie und Statistik **182** (1969), 398–433.
103. KORTE, B. AND VYGEN, J. : *Combinatorial optimization. Theory and algorithms*, Springer-Verlag, Berlin 2002.
104. KUO, C.C., GLOVER, F. AND DHIR, K.S.: *Analyzing and Modeling the Maximum Diversity Problem by Zero-One Programming*, Decision Sciences **24(6)** (1993), 1171-1185.
105. LAGUNA, M. AND MARTÍ, R.: *Scatter search: Methodology and implementations in C*, Kluwer Academic Publishers, 2003.
106. LAGUNA, M., MARTÍ, R. AND CAMPOS, V.: *Intensification and diversification with elite tabu search solutions for the linear ordering problem*, Computers and Operations Research **26** (1999), 1217–1230.
107. LENSTRA, H.W. JR.: *The acyclic subgraph problem*, Report BW26, Mathematisch Centrum, Amsterdam, 1973.
108. LEONTIEF, W.: *Quantitative input-output relations in the economic system of the United States*, The Review of Economics and Statistics **18** (1936).
109. LEONTIEF, W.: *Input-output economics*, Oxford University Press, 1966.
110. LEUNG, J. AND LEE, J.: *Reinforcing old fences gives new facets*, Tech. Rep 90-22, Dptm. of Operations Research, Yale University, 1990.
111. LEUNG, J. AND LEE, J.: *More facets from fences for linear ordering and acyclic subgraph polytopes*, Discrete Applied Mathematics **50** (1994), 185–200.
112. LOZANO, M., GLOVER, F., GARCÍA-MARTINEZ, C., RODRÍGUEZ, F. AND MARTÍ, R. : *Tabu Search with SO for the Quadratic Mininum Spanning Tree*, IIE Transactions **46 (4)** (2014) 414-428.
113. MARTÍ, R., DUARTE, A. AND REINELT, G.: *LOLIB. A library of benchmark instances for linear ordering problem*, 2010.
     https://www.uv.es/rmarti/paper/lop.html
114. MARTÍ, R. DUARTE, A., MARTÍNEZ-GAVARA, A. AND SÁNCHEZ-ORO, J.: *MDPLIB 2.0 - Maximum Diversity Problem Library*, 2021.
     https://www.uv.es/rmarti/paper/mdp.html
115. MARTÍ, R., GALLEGO, M. AND DUARTE, A.: *A Branch and Bound Algorithm for the Maximum Diversity Problem*, European Journal of Operational Research **200**(1), (2010), 36-44.
116. MARTÍ, R., GALLEGO, M., DUARTE, A. AND PARDO, E.: *Heuristics and Metaheuristics for the maximum diversity problem*, Journal of Heuristics **19**(4) (2013), 591-615.

117. MARTÍ, R., REINELT, G. AND DUARTE, A.: *LOLIB - Linear Ordering Problem Library*, https://www.uv.es/rmarti/paper/lop.html, 2010.

118. MARTÍ, R., LAGUNA, M., GLOVER, F. AND CAMPOS, V.: *Reducing the bandwidth of a sparse matrix with tabu search*, European J. of Operational Research **135** (2001), 450–459.

119. MARTÍ, R., MARTÍNEZ-GAVARA, A., PÉREZ-PELÓ, S. AND SÁNCHEZ-ORO, J.: *A review on diversity and dispersion maximization*, European J. of Operational Research (2021), forthcoming.

120. MARTÍ, R.: *Multi-start methods*, in: Glover, F. and Kochenberger, G. (eds.) *Handbook of Metaheuristics*, Kluwer Academic Publishers, 2003, 355–368.

121. MARTÍ, R., MARTÍNEZ-GAVARA, A. AND SÁNCHEZ-ORO, J.: *The Capacitated Dispersion Problem: A mathematical model and a scatter search metaheuristic*, Memetic Computing **13**, (2021) 131-146.

122. MARTÍNEZ-GAVARA, A., CORBERÁN, T. AND MARTÍ, R.: *GRASP and Tabu Search for the Generalized Dispersion Problem*, Expert Systems with Applications (2021), forthcoming.

123. MAYNE, D.Q. AND MEEWELLA, C.C.: *A non-clustering multistart algorithm for global optimization*; in: Bensoussan, A. and Lions, J.L. (eds.) *Analysis and Optimization of Systems*, Lecture Notes in control and information sciences 111, Springer, 1988, 334–345.

124. METROPOLIS, N., ROSENBLUTH, A. ROSENBLUTH, M., TELLER, A. AND TELLER, E.: *Equation of state calculations by fast computing machines*, J. of Chem. Physics **21** (1953), 1087–1092.

125. MICHALEWICZ, Z.: *Genetic algorithms + data structures = evolution programs*, Springer, 1994.

126. MITCHELL, J.E. AND BORCHERS, B: *Solving real-world linear ordering problems using a primal-dual interioir point cutting plane method*, Annals of Operations Research **62** (1996), 253–276.

127. MITCHELL, J.E. AND BORCHERS, B: *Solving linear ordering problems*, in: Frenk, H., Roos, K., Terlaky, T. and Zhang, s. (eds.), *High Performance Optimization*, Applied Optimization Vol. 33) Kluwer, 1999, 349–366 www.rpi.edu/~mitchj/generators/linord.

128. NADDEF, D. AND G. RINALDI, G.: *The crown inequalities for the symmetric traveling salesman polytope*, Mathematics of Operations Research **17** (1992), 308–326.

129. NEMHAUSER, G.L. AND WOLSEY, L.A.: *Integer and Combinatorial Optimization*, John Wiley and Sons, New York 1999.

130. NEWMAN, A.: *Cuts and orderings: On semidefinite relaxations for the linear ordering problem*, in: *Proc. of APPROX 2004*, LNCS 3122, Springer, 195–206.

131. NERI, F. AND COTTA., C.: *A Primer on Memetic Algorithms, Handbook of Memetic Algorithms*, in: Neri, F., Cotta c. and Moscato, P. (eds.) *Studies in Computational Intelligence*, 379, Springer, 2012, 43-54.

132. NEWMAN, A. AND VEMPALA, S.: *Fences are futile: On relaxations for the linear ordering problem*, in: *Proc. of the 8th Int. IPCO Conference*, LNCS 2081, Springer, 2001, 333–347.

133. ONWUBOLU, G.C. AND BABU, B.V.: *New optimization techniques in engineering*, Studies in Fuzziness and Soft Computing, Vol. 141, Springer, 1994.

134. OSMAN, I.H. AND KELLY, J.P.: *Meta-heuristics: Theory and applications*, Kluwer Academic Publishers, 1996.

135. OSWALD, M., REINELT, G. AND SEITZ, H.: *Applying mod-k-cuts for solving linear ordering problems*, Top **17** (2009), 158–170.

136. PAGE, S.E.: *The difference ? How the power of diversity creates better groups, firms, schools, and societies*, Princeton University Press, 2007.

137. PALUBECKIS, G.: *Iterated tabu search for the maximum diversity problem*, Applied Mathematics and Computing **189**(1), (2007), 371-383.

138. PARREÑO, F., ÁLVAREZ-VALDÉS, R. AND MARTÍ, R.: *Measuring Diversity. A review and an empirical analysis*, European Journal of Operation Research **289**, (2021), 515-532.

139. PROKOPYEV O.A,, KONG, N. AND MARTINEZ-TORRES, D.L.: *The equitable dispersion problem*, European Journal of Operation Research **197**, (2009), 59-67.

140. POETSCH, G.: *Lösungsverfahren zur Triangulation von Input-Output-Tabellen*, Rechenzentrum Graz, Heft 79, 1973.
141. RALPHS, T.K. AND LADANYI, L.: *SYMPHONY: A parallel framework for branch and cut*, Rice University, 2000.
     http://branchandcut.org
142. REEVES, C.: *Modern heuristics techniques for combinatorial optimization problems*, Advanced topics in computer science, Mc Graw-Hill, 1995.
143. REINELT, G.: *The linear ordering problem: Algorithms and applications*, Research and exposition in mathematics **8**, Heldermann, 1985.
144. REINELT, G.: *A note on small linear ordering polytopes*, Discrete & Computational Geometry **10** (1993), 67–78.
145. RENDL, F.: *Semidefinite relaxations for integer programming*, in: Jünger, M., Liebling, Th.M., Naddef, D., Nemhauser, G.L., Pulleyblank, W.R., Reinelt, G., Rinaldi, G., Wolsey, L.A. (eds.) *50 Years of Integer Programming 1958-2008: From the Early Years to the State-of-the-Art*, Springer, 2009, 687–726.
146. RESENDE, M.G.C., MARTÍ, R., GALLEGO, M. AND DUARTE, A.: *GRASP and path relinking for the max−min diversity problem*, Computers and Operations Research **37(3)**, (2010), 498-508.
147. RIBEIRO, C.C., UCHOA, E. AND WERNECK, R.F.: *A hybrid GRASP with perturbations for the Steiner problem in graphs*, INFORMS Journal on Computing **14** (2002), 228 - 246.
148. RINNOOY KAN, A.H.G. AND TIMMER, G.T.: *Global optimization*, in: Nemhauser, G.L., Rinnooy Kan, A.H.G. and Todd, M.J. (eds.) *Handbooks in operations research and management science, Vol. 1*, North Holland, 1989, 631–662.
149. ROSENKRANTZ, D.J., TAYI, G.K. AND RAVI, S.S.: *Facility Dispersion Problems under Capacity and Cost Constraints*, Journal of Combinatorial Optimization **4** (2000), 7–33.
150. RUIZ, R. AND STÜTZLE, T.: *A simple and effective iterated greedy algorithm for the permutation flowshop scheduling problem*, European Journal of Operational Research **177 (3)** (2008), 2033–2049.
151. SANDOYA, F., ACEVES, R., MARTÍNEZ-GAVARA, A., DUARTE, A., AND MARTÍ, R.: *Diversity and Equity Models*, in: Martí, R., Resende, M., Pardalos, P. (eds.) *Handbook of Heuristics*, Springer (2018) 979-998.
152. SANTUCCI,V. AND CEBERIO, J.: *Using Pairwise Precedences for Solving the Linear Ordering Problem*, Applied Soft Computing **87** (2020).
153. SILVA, G.C., OCHI, L.S. AND MARTINS, S.L.: *Experimental comparison of greedy randomized adaptive search procedures for the maximum diversity problem*, Lecture Notes in Computer Science **3059**, (2004), 498-512, Springer.
154. SCHIAVINOTTO, T. AND STÜTZLE, T.: *The linear ordering problem: Instances, search space analysis and algorithms*, J. of Mathematical Modelling and algorithms **3** (2004), 367–402.
155. SCHRIJVER, A.: *Theory of linear and integer programming*, John Wiley and Sons, New York 1999.
156. SCHRIJVER, A.: *Combinatorial optimization. Polyhedra and efficiency*, Springer-Verlag, Berlin 2003.
157. SLATER, P.: *Inconsistencies in a schedule of paired comparisons*, Biometrika **48** (1961), 303–312.
158. SOLIS, F. AND WETS, R.: *Minimization by random search techniques*, Math. of Operations Research **6** 1981, 19–30.
159. SUCK, R.: *Geometric and combinatorial properties of the polytope of binary choice probabilities*, Math. Social Sciences **23** (1992), 81–102.
160. TRANSPARENCY INTERNATIONAL:
     www.transparency.org/policy_research/surveys_indices/cpi.
161. TSPLIB: *A library of benchmark instances for the traveling salesman problem*,
     www.ifi.uni-heidelberg.de/groups/comopt/software/TSPLIB95.

162. WANG, Y., HAO, J.-K., GLOVER, F. AND LÜ, Z.: *A tabu search based memetic algorithm for the maximum diversity problem*, Engineering Applications of Artificial Intelligence **27** (2014) 103–114.
163. WOLSEY, L.A. *Integer Programming*, John Wiley and Sons, New York 1998.
164. YOUNG, H. P.: *On permutations and permutation polytopes*, Mathematical Programming Study **8** 1978, 128–140.
165. ZHOU, Y., AND HAO, J.K. AND DUVAL, B.: *Opposition-based memetic search for the maximum diversity problem*, IEEE Transactions on Evolutionary Computation **5** 2017, 731–745.

Index

$\angle_m$-inequalities, 175
$\mathcal{H}$-polyhedron, 167
$\mathcal{V}$-polyhedron, 168
σ-class, 177
ε-approximative, 31
$\{0, \frac{1}{2}\}$-Chvátal-Gomory inequality, 155
k-fence, 172
k-fence inequality, 172
k-th unit vector, 167
t-reinforced k-fences, 175
P_{LO}^n, 169
P_{LO}^n-class, 179
0/1-IP, 144

active node, 156
active variable, 156
acyclic subdigraph polytope, 181
acyclic subdigraph problem, 3
acyclic tournament, 2
adaptive, 75
adaptive multi-start heuristic (AMS), 47
affine combination, 167
affine hull, 167
affine rank, 167
affine space, 167
aggregation of individual preferences, 4
approximative algorithm, 27
arc reversal lemma, 170
attribute, 75
attributive memory, 75
augmented k-fences, 175

benchmark problems, 10, 15, 119, 120
best-first, 159
binary choice probabilities, 5
Boltzmann constant, 84
bounding, 127

branch-and-bound tree, 126
branch-and-cut method, 125
branching, 127
breadth-first, 159
bundle method, 131

canonical LP relaxation, 144
careful annealing, 83
characteristic vector, 144, 169
chromosome reconstruction problem, 85
Chvátal rank, 153
closure, 153
conic combination, 167
conic hull, 167
convex combination, 167
convex hull, 167
corruption perception index, 7
coupled task problem, 9
crossing minimization problem, 8
crossover, 110
current node, 156
cut-width problem, 10
cutting plane approach, 146

degree of linearity, 11, 196
depth-first, 158
diagonal inequalites, 175
dicycle relaxation, 182, 207
dicycle strength, 182
dimension, 168
dissimilarity, 53
diversification, 51, 76
diversification generation method, 95
diversity, 53
dynamic program, 125

elementary closure, 153

© Springer-Verlag GmbH Germany, part of Springer Nature 2022
R. Martí and G. Reinelt, *Exact and Heuristic Methods in
Combinatorial Optimization*, Applied Mathematical Sciences 175,
https://doi.org/10.1007/978-3-662-64877-3

Printed in the United States
by Baker & Taylor Publisher Services

Printed in the United States
by Baker & Taylor Publisher Services